W0227646

A CATALOGUE

of

BRITISH PLANTS,

ARRANGED ACCORDING TO THE NATURAL SYSTEM,

WITH THE SYNONYMS

OF

DE CANDOLLE, SMITH, AND LINDLEY.

BY THE

Rev. J. S. HENSLOW, M.A.

PROFESSOR OF BOTANY

IN THE UNIVERSITY OF CAMBRIDGE.

Cambridge:

Printed by James Hodson, Trinity Street.

SOLD BY J. AND J. J. DEIGHTON, AND T. STEVENSON, CAMBRIDGE;

AND C. J. G. AND F. RIVINGTON, AND HATCHARD

AND SON, LONDON.

1829

This Catalogue being intended chiefly for the use of those Students who attend the Botanical Lectures in the University of Cambridge, all the "Species" not hitherto met with in the County have been printed in *Italics*, to mark the extent of our local flora.

It may be made serviceable during a Botanical excursion in any distant part of the kingdom, by preventing the traveller from burdening himself unnecessarily by drying those plants which he might procure nearer home. A third use which this Catalogue may be made to serve, is for a list upon which to mark off the several species as they are placed in the herbarium. If interleaved it may be useful for registering, during the year, the times of flowering or peculiar habitats of the rarer species.

E R R A T A.

Page 3. After SENEBIERA *didyma* add,
                              (pinnatifida. D. C.)
  5. For STAPHYLEÆ, read STAPHYLEA.
  10. SAXIFRAGA 7. " oppositifolia" should be in *Italics*.
  15. For *Empetrum* nigrum *read* EMPETRUM *nigrum*.
  24. For SCHÆNUS, read SCHŒNUS.
  26. For PHLEUM, 9. *read* PHLEUM, 6.

Natural Arrangement of British Plants.

Note.—The species in *Italics* are not found in Cambridgeshire
  * Species probably only *naturalized* in Britain.
  D. C. De Candolle, Botanicon Gallicum ; & Prodromus.
  Sm. Smith's English Flora.
  Lind. Lindley's Synopsis of the British Flora.

DICOTYLEDONES
OR
EXOGENÆ.

I. RANUNCULACEÆ.
1. CLEMATIS.
   1. vitalba.
2. THALICTRUM.
   1. *alpinum.*
   2. minus.
   3. *majus.*
   4. flavum.
3. ANEMONE.
   1. pulsatilla.
   2. *apennina.*
   3. nemorosa.
   4. *ranunculoides.*
4. ADONIS.
   1. *autumnalis.*
5. MYOSURUS.
   1. minimus.
6. RANUNCULUS.
   1. hederaceus.
   2. aquatilis. *Sm.*
      α. heterophyllus.
      β. pantothrix.
      γ. circinatus.
      δ. fluviatilis.
   3. *alpestris.*
   4. *gramineus.*
   5. lingua.
   6. flammula.
      β. *reptans.*
   7. auricomus.
   8. sceleratus.
   9. acris.
   10. repens.
   11. bulbosus.

12. hirsutus.
    (*philonotis.* *D.C.*)
    β. *parvulus.*
13. arvensis.
14. parviflorus.
15. ficaria.
    (*Ficaria
       ranunculoides.* *D. C.*)
7. CALTHA
   1. palustris.
      β. *minor.*
   2. *radicans.*
8. TROLLIUS.
   1. *europæus.*
9. HELLEBORUS
   1. viridis.
   2. fœtidus.
10. AQUILEGIA.
    1. vulgaris.
11. DELPHINIUM.
    *1. consolida.
12. ACONITUM.
    *1. *napellus.*
13. ACTÆA.
    1. *spicata.*
14. PÆONIA.
    *1. *corallina.*

II. BERBERIDEÆ.
1. BERBERIS.
   1. vulgaris.
2. EPIMEDIUM.
   *1. *alpinum.*

III. NYMPHÆACEÆ.
1. NYMPHÆA.
   1. alba.
2. NUPHAR.
   1. lutea.
   2. *pumila.*

A

IV. PAPAVERACEÆ.
1. PAPAVER.
  1. hybridum.
  2. argemone.
  β. maritimum.
  3. nudicaule.
  4. dubium.
  5. rhœas.
  *6. somniferum.
2. MECONOPSIS.
  (Papaver. Sm.)
  1. cambrica.
3. RŒMERIA.
  (Glaucium. Sm.)
  *1. violacea.
  (hybrida D. C.)
4. GLAUCIUM.
  1. luteum.
  (flavum D. C.)
  *2. phœniceum.
  (corniculatum. D. C.)
5. CHELIDONIUM.
  1. majus.
  β. laciniatum. Sm.
V. FUMARIACEÆ.
1. CORYDALIS.
  (Fumaria. Sm.)
  1. solida.
  (bulbosa. D. C.)
  2. lutea.
  3. claviculata.
2. FUMARIA.
  1. officinalis.
  2 ? capreolata.
  3. parviflora.
VI. CRUCIFERÆ.
1. MATTHIOLA.
  1. incana.
  2. sinuata.
2. CHEIRANTHUS.
  1. cheiri.
  (fruticulosus. Sm.)
3. NASTURTIUM.
  1. officinale.
  2. sylvestre.
  3. terrestre.
  (palustre D. C.)
  4. amphibium.

4. BARBAREA.
  1. vulgaris.
  2. præcox.
5. TURRITIS.
  1. glabra.
6. ARABIS.
  1. hirsuta.
  2. stricta.
  3. ciliata.
  4. thaliana.
  5. hispida.
  6. turrita.
7. CARDAMINE.
  1. bellidifolia.
  2. amara.
  3. pratensis.
  4. hirsuta.
  5. impatiens.
8. DENTARIA.
  1. bulbifera.
9. ALYSSUM.
  (Glyce. Lind.)
  1. maritimum.
10. DRABA.
  1. aizoides.
  2. hirta.
  (rupestris D. C.)
  3. incana.
  (contorta D. C.)
  4. muralis.
11. EROPHILA.
  1. vulgaris.
  (Draba verna. Sm.)
12. COCHLEARIA.
  1. armoracia.
  2. anglica.
  3. officinalis.
  4. grœnlandica.
  5. danica.
  β. integrifolia.
13. THLASPI.
  1. arvense.
  2. perfoliatum.
  3. alpestre.
14. HUTCHINSIA.
  1. petræa.
15. TEESDALIA.
  1. nudicaulis.

16. IBERIS.
 1. *amara.*
17. CAKILE.
 1. *maritima.*
18. HESPERIS.
 1. *matronalis.*
19. SISYMBRIUM.
 1. officinale.
 2. irio.
 3. sophia.
20. ALLIARIA.
 1. officinalis.
 (Erysimum alliar: Sm.)
21. ERYSIMUM.
 1. cheiranthoides.
 2. *orientale.*
 (perfoliatum   D. C.)
22. CAMELINA.
 *1. sativa.
23. SENEBIERA.
 (Coronopus   Lind.)
 1. *didyma.*
 2. coronopus.
 (C. ruellii   Lind.)
24. CAPSELLA.
 (Thlaspi.   Sm.)
 1. bursa-pastoris.
25. LEPIDIUM.
 1. campestre.
  α. vulgatius.
  β. *glabrum.*
  γ. *hirtum.*
 2. *hirtum.*
 3. ruderale.
 4. latifolium.
26. ISATIS.
 *1. tinctoria.
27. BRASSICA.
 1. *oleracea.*
 2. *campestris.*
 3. rapa.
 4. napus.
 5. *monensis.*
28. SINAPIS.
 1. nigra.
 2. arvensis.
 3. alba.

29. DIPLOTAXIS.
 (Sinapis.   Sm.)
 1. *tenuifolia.*
 2. *muralis.*
30. CARRICHTERA.
 *1. *vellæ.*
 (Vella annua.   Sm.)
31. CRAMBE.
 1. *maritima.*
32. RAPHANUS.
 1. raphanistrum.
 2. *maritimus.*
33. SUBULARIA.
 1. *aquatica.*
VII. CISTINEÆ.
 1. HELIANTHEMUM.
 (Cistus.   Sm.)
 1. *guttatum.*
 2. *ledifolium.*
 3. *marifolium.*
 4. vulgare.
 (C. helianth: Sm.)
 β. *tomentosum.*
 5. *surrejanum.*
 6. *polifolium.*
VIII. VIOLARIEÆ.
 1. VIOLA.
 1. *palustris.*
 2. hirta.
 3. odorata.
 4. canina.
 5? flavicornis.
 6. lactea.
 (montana   γ.   D.C.)
 7. *lutea.*
 (sudetica   α.   D. C.)
 β. *amœna.*
 8. tricolor.
 β. arvensis.
IX. RESEDACEÆ.
 1. RESEDA.
 1. lutea.
 2. luteola.
X. DROSERACEÆ.
 1. DROSERA.
 1. rotundifolia.
 2. longifolia

3? anglica.

15. CERASTIUM.
1. vulgatum.
2. viscosum.
3. semidecandrum.
  β. pumilum.
  γ. tetrandrum. Lind.
4. aquaticum
5. alpinum.
6. latifolium. Lind.
7. arvense.
16. CHERLERIA.
1. sedoides.
XIV. LINEÆ.
1. LINUM.
  *1. usitatissimum.
  2. angustifolium.
  3. perenne.
    β. procumbens.
  4. catharticum.
2. RADIOLA.
  1. millegrana.
XV. MALVACEÆ.
1. MALVA.
  1. moschata.
  2. sylvestris.
  3. rotundifolia.
    β. pusilla.
2. ALTHÆA.
  1. officinalis.
  *2. hirsuta.
3. LAVATERA.
  1. arborea.
XVI. TILIACEÆ.
1. TILIA.
  *1. europæa.
    (intermedia D C.)
  *2. grandifolia.
    (platyphylla D.C.)
  *β. corallina. Sm.
    (T. rubra D.C.)
  3. parvifolia.
XVII. HYPERICINEÆ.
1. HYPERICUM.
  1. androsæmum.
    (Androsæmum
    officinale. D. C.)
  2. calycinum.

3. quadrangulum.
  α. vulgare.
  β. dubium. D. C.
4. humifusum.
5. perforatum.
6. elodes.
7. hirsutum.
8. pulchrum.
9. barbatum.
10. montanum.
XVIII. ACERINEÆ.
1. ACER.
  *1. pseudo-platanus.
  2. campestre.
XIX. GERANIACEÆ.
1. GERANIUM.
  1. sanguineum.
    β. prostratum.
  2. nodosum.
  3. phœum.
  4. sylvaticum.
  5. pratense.
  6. pyrenaicum.
  7. molle.
  8. pusillum.
    β. humile.
  9. rotundifolium.
  10. columbinum.
  11. dissectum.
  12. lucidum.
  13. robertianum.
    β. raii. Sm.
2. ERODIUM.
  1. cicutarium.
  2. moschatum.
  3. maritimum.
XX. BALSAMINEÆ.
1. IMPATIENS.
  1. noli-me-tangere.
XXI. OXALIDEÆ.
1. OXALIS.
  1. corniculata.
  2. acetosella.
XXII. CELASTRINEÆ.
  (STAPHYLIACEÆ. Lind.)
1. STAPHYLEÆ.
  1. pinnata.

(*CELAST.VERÆ. Lind.*)
2. EUONYMUS.
1. europæus.

(*ILICINEÆ, Lind.*)
3. ILEX.
1. aquifolium.
XXIII. RHAMNEÆ.
1. RHAMNUS.
1. catharticus.
2 frangula.
XXIV. LEGUMINOSÆ.
1. ULEX.
1. europæus.
2. nanus.
2. GENISTA.
1. anglica.
2. tinctoria.
3. *pilosa.*
3. CYTISUS.
(*Spartium. Sm.*)
1. scoparius.
4. ONONIS.
1. arvensis. *Sm.*
α. inermis.
β. spinosa.
γ. *repens.*
(*O. procurrens D. C.*)
5. ANTHYLLIS.
1. vulneraria.
α. vulgaris *D. C.*
β. *dillenii.*
(*rubriflora D. C.*)
6. MEDICAGO.
1. lupulina.
β. *polycarpos Sm.*
2. falcata.
*3. sativa.
4. minima.
5. maculata.
6. *muricata.*
7. TRIGONELLA.
(*Trifolium. Sm.*)
1. *ornithopodioides.*
8. MELILOTUS.
(*Trifolium. Sm.*)
1. officinalis.

9. TRIFOLIUM.
1. arvense.
2. striatum.
3. scabrum.
4. maritimum.
5. ochroleucum.
6. medium.
7. pratense.
8. *stellatum.*
9. *suffocatum.*
10. *glomeratum.*
11. repens.
12. fragiferum.
13. subterraneum.
14. procumbens.
15. minus.
16? filiforme.
10. LOTUS.
1. corniculatus.
β. minor. *Sm.*
γ. major. *D. C.*
2. *decumbens.*
3. *angustissimus.*
11. OXYTROPIS.
(*Astragalus. Sm.*)
1. *uralensis.*
2. campestris.
12. ASTRAGALUS.
1. hypoglottis.
2. glycyphyllus.
13. ORNITHOPUS.
1. perpusillus.
14 HIPPOCREPIS.
1. comosa.
15. ONOBRYCHIS.
1. sativa.
(*Hedysarum onob: Sm.*)
16. VICIA.
1. sylvatica.
2. cracca.
3. sativa.
α. vulgaris.
β. sylvestris.
4. *angustifolia.*
5. *lathyroides.*
6. *lævigata.*
7. *lutea.*
8. *hybrida.*

9. sepium.
17. ERVUM.
  1. hirsutum.
  2. tetraspermum.
18. PISUM.
  1. *maritimum.*
19. LATHYRUS.
  1. sylvestris.
  2. latifolius.
  3. pratensis.
  4. palustris.
  5. aphaca.
  6. nissolia.
  7. *hirsutus.*
  8. *bithynicus.*
    (*Vicia bith : Sm.*)
20. OROBUS.
  1. *niger.*
  2. *sylvaticus.*
    β. *simplicifolius. Sm.*
  3. *tuberosus.*
    β. *tenuifolius.*
XXV. ROSACEÆ.
  1. PRUNUS.
  1. spinosa.
  2? insititia *D. C.*
  *3. domestica.
  2. CERASUS.
    (*Prunus. Sm.*)
  1. avium.
    (*P. cerasus Sm.*)
  2. padus.
  3. SPIRÆA.
  1. *salicifolia.*
  2. ulmaria.
  3. filipendula.
  4. DRYAS.
  1. *octopetala.*
  5. GEUM.
  1. urbanum.
    α. vulgare.
    β. sylvestre.
  2. rivale.
    β. hybridum.
  6. RUBUS.
    (*Smith.*)
  1. fruticosus
  2. *plicatus*

3. *rhamnifolius*
  β. *cordifolius*
4. *leucostachys*
5. *glaudulosus*
6. *nitidus*
7. *affinis*
8. *suberectus*
9. idæus
10. corylifolius
11. cæsius
12. *saxatilis*
13. *arcticus*
14. *chamæmorus*

(*Lindley.*)
1. *plicatus.*
2. *fastigiatus.*
3. *suberectus.*
4. *affinis.*
5. *nitidus.*
6. *cordifolius.*
7. *rhamnifolius.*
8. fruticosus.
9. *abruptus.*
10. *saxatilis.*
11. *macrophyllus.*
12. *vulgaris.*
13. *leucostachys.*
14. *diversifolius.*
15. *discolor.*
16. *fusco-ater.*
17. *pallidus.*
18. *köhleri.*
19. *rudis.*
20. *echinatus.*
21. *hirtus.*
22. *dumetorum.*
23. cæsius.
24. idæus.
25. *arcticus.*
26. *chamæmorus.*
7. FRAGARIA.
  1. vesca.
  2? elatior.
    (*moschata. Lind.*)
  3. *calycina.*
8. POTENTILLA.
  1. tormentilla *D. C.*
(*Tormentilla officinalis. Sm.*)

β. *nemoralis* **D. C.**
(*Tor. reptans. Sm.*)
2. reptans.
3. verna.
4. *alpestris.*
5. argentea.
6. *opaca.*
7. *fruticosa.*
8. anserina.
9. comarum.
(*Comarum palustre. Sm.*)
10. *rupestris.*
11. *alba.*
12. *tridentata.*
13. fragariastrum.
(*fragaria. D.C.*)
9. SIBBALDIA.
1. *procumbens.*
10. AGRIMONIA.
1. eupatoria.
11. ALCHEMILLA.
1. vulgaris.
2. *alpina.*
3. arvensis.
12. SANGUISORBA.
1. officinalis.
2? *media.*
13. POTERIUM.
1. sanguisorba.
14. ROSA.
(*Smith.*)
*1. *cinnamomea*
2. *rubella*
3. spinosissima
β. *ciphiana*
γ. *pusilla*
4. *involuta*
5. *doniana*
6. *gracilis*
7. *sabini*
8. villosa
β. *heterophylla*
γ. *pulchella*
9. tomentosa
β. *scabriuscula*
10. *subglobosa*
11. rubiginosa
12. *micrantha*

13. *borreri*
14. *cæsia*
15. *sarmentacea*
16. *bractescens*
17. *dumetorum*
18. *forsteri*
19. *hibernica*
20. canina
β. *surculosa*
γ. *nuda*
21. *systyla*
β. *lanceolata*
γ. *monsoniæ*
22. arvensis

(*Lindley.*)
1. *dicksoniana*
2. *rubella*
3. spinosissima
β. *pilosa*
4. *hibernica*
5. *involuta*
6. *sabini*
β. *doniana*
7. tomentosa
β. *fœtida*
8. mollis
β. *resinosa*
9. *sylvestris*
10. rubiginosa
β. *micrantha*
γ. *umbellata*
δ. *inodora*
† *parvifolia*
11. *sepium*
12. canina
β. *sarmentacea*
13. *collina*
14. *dumetorum*
15. *cæsia*
16. *systyla*
α. *ovata*
β. *lanceolata*
17. arvensis

(*POMACEÆ. Lind.*)
15. CRATÆGUS.

(*Mespilus. Sm.*)
1. oxyacantha.
  β. eriocarpa.
16. COTONEASTER.
  1. *vulgaris.*
  (*Mespilus cotoneaster. Sm.*)
17. MESPILUS.
  1. *germanica.*
18. PYRUS.
  1. communis.
  2. malus.
  3 *aria.*
    β. *intermedia. Sm.*
  4? *pinnatifida.*
  5. torminalis.
  6. aucuparia.
  7. *domestica.*

XXVI. CUCURBITACEÆ.
  1. BRYONIA.
  1. dioica.
XXVII. ONAGRARIÆ.
  1. EPILOBIUM.
    1. *angustifolium.*
    2. *alpinum.*
    3. *alsinifolium.*
      (*origanifolium. D. C.*)
    4. montanum.
    5. hirsutum.
    6. palustre.
    7. parviflorum.
      (*molle. D. C.*)
    8. *roseum.*
    9. tetragonum.
  2. ŒNOTHERA.
  *1. *biennis.*
  3. ISNARDIA.
  1. *palustris.*

  (*CIRCÆACEÆ. Lind.*)
  4. CIRCÆA.
  1. lutetiana.
  2? *alpina.*
    β. *intermedia.*

XXVIII. HALORAGEÆ.
  1. MYRIOPHYLLUM.
  1. spicatum.
  2. verticillatum.

(*CALLITRICHINEÆ. Lind.*)
2. CALLITRICHE.
  1. aquatica.
  α. verna.
  β? autumnalis.
XXIX. HIPPURIDEÆ.
  1. HIPPURIS.
  1. vulgaris.
XXX. CERATOPHYLLEÆ.
  1. CERATOPHYLLUM.
  1. demersum.
  2? *submersum.*
XXXI. LYTHRARIEÆ.
  (*SALICARIÆ. Lind.*)
  1. LYTHRUM.
  1. hyssopifolium.
  2. salicaria.
  2. PEPLIS.
  1. portula.
XXXII. TAMARISCINEÆ.
  1. TAMARIX.
  1. *gallica.*
XXXIII. PORTULACEÆ.
  1. MONTIA.
  1. fontana.
XXXIV. PARONYCHIEÆ.
  1. CORRIGIOLA.
  1. *littoralis.*
  2. HERNIARIA.
  1. *glabra.*
  2? *hirsuta.*
  3. ILLECEBRUM.
  1. *verticillatum.*
  4. POLYCARPON.
  1. *tetraphyllum.*

  (*SCLERANTHEÆ. Lind.*)
5. SCLERANTHUS.
  1. annuus.
  2. perennis.
XXXV. CRASSULACEÆ.
  1. TILLÆA.
  1. *muscosa.*
  2. COTYLEDON.
    (*Umbilicus. D. C.*)
  1. *umbilicus.*
    (*U. pendulinus. D. C.*)
  2. *lutea.*
    (*U. erectus. D. C.*)

B

3. **RHODIOLA.**
   1. *rosea.*
   (*Sedum rhod:* D. C.)
4. **SEDUM.**
   1. telephium.
   2. *anglicum.*
   3. *villosum.*
   4. dasyphyllum.
   5. album.
   6. acre.
   7. sexangulare.
   8. *rupestre.*
   9. *forsterianum.*
   10. *glaucum.* Sm.
   (*albescens.* Lind.)
   11. reflexum.
5. **SEMPERVIVUM.**
   *1. tectorum.

XXXVI. GROSSULARIEÆ
1. **RIBES.**
   *1. grossularia.
   *β. uva-crispa.
   2. *alpinum.*
   3. *spicatum.*
   4. rubrum.
   5. *petræum.*
   6. nigrum.

XXXVII. SAXIFRAGEÆ.
1. **SAXIFRAGA.**
   (*Hirculus.* Lind.)
   1. *hirculus.*
   (*H. ranunculoides.* Lind.)

   (*Leiogyne.* Lind.)
   2. *aizoides.*
   3. granulata.
   4. *cernua.*
   5. *rivularis.*
   6. *nivalis.*

   (*Saxif. veræ.* Lind.)
   7. oppositifolia.
   8. tridactylites.
   9. *muscoides.*
   10. *pygmæa.*
   11. *cæspitosa.*
   12. *hirta.*
   13. *affinis.*

14. *platypetala.*
15. *incurvifolia.*
16. *denudata.*
17. *hypnoides.*
   β. *condensata.* Lind.
   γ. *elongella.* Lind.
18. *leptophylla.*
19. *lætevireñs.*
20. *pedatifida.*

   (*Robertsonia.* Lind.)
21. *geum.*
   β. *elegans.*
   γ. *guttata.*
22. *hirsuta.*
   β. *depilata.*
23. *umbrosa.*
   β. *punctata.*
   γ. *serratifolia.*
24. *stellaris.*

2. **CHRYSOSPLENIUM.**
   1. *alternifolium.*
   2? *oppositifolium.*
3. **ADOXA.**
   1. moschatellina.

XXXVIII. UMBELLIFERÆ
1. **DAUCUS.**
   1. *maritimus.*
   2. carota.
2. **CAUCALIS.**
   1. daucoides.
   *2. latifolia.
3. **TORILIS.**
   1. nodosa.
   2. anthriscus.
   3. infesta.
4. **CORIANDRUM.**
   *1. sativum.
5. **TORDYLIUM.**
   *1. *officinale.*
   (*Condylocarpus off:* Lind.)
   2. *maximum.*
6. **HERACLEUM.**
   1. spondylium.
   α. vulgare.
   β. *angustifolium.*

7. PASTINACA.
1. sativa.
8. PEUCEDANUM.
1. officinale.
2. palustre.
(*Selinum pal: Sm.*)
9. IMPERATORIA.
(*Peucedanum. Lind.*)
1. *ostruthium.*
10. ARCHANGELICA.
*1. *officinalis.*
(*Angelica arch. Sm.*)
11. ANGELICA.
1. sylvestris.
12. BUPLEURUM.
1. tenuissimum.
2. *odontites.*
3. rotundifolium.
13. BUNIUM.
(*Conopodium. Lind.*)
1. flexuosum.
14. PIMPINELLA.
1. magna.
2. saxifraga.
15. SIUM.
1. latifolium.
2. angustifolium.
16. LIGUSTICUM.
1. *scoticum.*
2. *cornubiense.*
(*Physospermum commutatum. Lind.*)
17. CNIDIUM.
1. silaus.
(*Silaus pratensis. Lind.*)
18. MEUM.
1. *athamanticum.*
19. ÆGOPODIUM.
1. podagraria.
20. CARUM.
*1. carui.
2. *verticillatum.*
(*Sium vertic: Sm.*)
21. PETROSELINUM.
*1. *sativum.*
2. segetum.
(*Sison seg: Sm.*)
22. APIUM.
1. graveolens.

23. ÆTHUSA.
1. cynapium.
24. TRINIA.
(*Pimpinella. Sm.*)
1. *dioica.*
(*glaberrima. Lind.*)
25. SISON.
1. amomum.
26. SESELI.
(*Athamanta. Sm.*)
1. libanotis.
27. HELOSCIADIUM.
(*Sium. Sm.*)
1. inundatum.
2. repens.
3. nodiflorum.
28. FŒNICULUM.
1. officinale.
(*vulgare. Lind.*)
(*Meum fœniculum. Sm.*)
29. ŒNANTHE.
1. phellandrium.
2. fistulosa.
3. pimpinelloides.
4. peucedanifolia.
5. *crocata.*
30. CICUTA.
1. virosa.
31. CRITHMUM.
1. *maritimum.*
32. CHÆROPHYLLUM.
(*Myrrhis. Sm.*)
1. *aromaticum.*
2. *aureum.*
3. temulentum.
(*temulum. D. C.*)
33. ANTHRISCUS.
1. sylvestris.
(*Chærophyllum syl. Sm*)
2. *cerefolium.*
(*Chær. sativum. Sm.*)
3. vulgaris.
34. SCANDIX.
1. pecten-veneris.
35. MYRRHIS.
1. *odorata.*
36. SMYRNIUM.
1. olusatrum.

37. CONIUM.
   1. maculatum.
38. ECHINOPHORA.
   1. spinosa.
39. SANICULA.
   1. europæa.
40. ERYNGIUM.
   1. campestre.
   2. maritimum.
41. HYDROCOTYTE.
   1. vulgaris.
XXXIX. CAPRIFOLIACEÆ
   1. HEDERA.
      1. helix.
   2. CORNUS.
      1. sanguinea.
      2. suecica.
   3. SAMBUCUS.
      1. ebulus.
      2. nigra.
         β. laciniata.
   4. VIBURNUM.
      1. lantana.
      2. opulus.
   5. LONICERA.
      (Caprifolium. Lind.)
      1. caprifolium.
      (C. perfoliatum. Lind.)
      2. periclymenum.
      ————
      (Lonic: veræ. Lind.)
      3. xylosteum.
   6. LINNÆA.
      1. borealis.
XL. LORANTHEÆ.
   1. VISCUM.
      1. album.
XLI. RUBIACEÆ.
   (STELLATÆ. Lind.)
   1. RUBIA.
      1. peregrina.
   2. GALIUM.
      1. cruciatum.
      2. palustre
      3? witheringii.
      4. saxatile.
      5. uliginosum.
      6. erectum.

7? cinereum.
8. aristatum.
9. verrucosum.
10. tricorne.
11. spurium.
12. pusillum.
13. verum.
14. mollugo.
15. anglicum.
16. boreale.
17. aparine.
   3. ASPERULA.
      1. odorata.
      2. cynanchica.
   4. SHERARDIA.
      1. arvensis.
XLII. VALERIANEÆ.
   1. FEDIA.
      (Valerianella D. C.)
      1. olitoria.
      2. dentata.
   2. CENTRANTHUS.
      (Valeriana. Sm.)
      *1. ruber.
      (latifolius. D. C.)
   3. VALERIANA.
      1. dioica.
      2. officinalis.
      3. pyrenaica.
XLIII. DIPSACEÆ.
   1. SCABIOSA.
      1. columbaria.
      2 succisa.
   2. KNAUTIA.
      (Scabiosa. Sm.)
      1. arvensis.
   3. DIPSACUS.
      *1. fullonum.
      2. sylvestris.
      3. pilosus.
XLIV. COMPOSITÆ.
   1. EUPATORIUM.
      1. cannabinum.
   2. TUSSILAGO.
      1. farfara.
      2. petasites.
         β. hybrida.
         (fœmina. Lind.)

3. CINERARIA.
 1. integrifolia.
  α. campestris.
  β. maritima.
 2. palustris.
4. SENECIO.
 1. jacobæa.
 2 ? aquaticus.
 3. tenuifolius.
 *4. squalidus.
 5. sylvaticus.
 6. viscosus.
 7. lividus.
 8. paludosus.
 9. saracenicus.
 10. vulgaris.
5. DORONICUM.
 1. pardalianches.
6. CHRYSOCOMA.
 1. linosyris.
7. ASTER.
 1. tripolium.
8. ERIGERON.
 *1. canadensis.
 2. acris.
 3. alpinus.
  β. uniflorus. Lind.
9. SOLIDAGO.
 1. virgaurea.
  α. aurea.
  β. cambrica.
10. BELLIS.
 1. perennis.
11. CONYZA.
 1. squarrosa.
12. INULA.
 1. helenium.
 2. crithmoides.
(Limbarda tricuspis. Lind.)

 (Pulicaria. Lind.)
 3. dysenterica.
 4. pulicaria.
  (P. vulgaris. Lind.)
13. GNAPHALIUM.
 1. luteo-album.
 2. sylvaticum. D. C.
  α. fuscum.

β. rectum.
3. supinum.
4. uliginosum.

 (Filago. Lind.)
5. gallicum.
6. germanicum.
7. minimum.

 (Antennaria. Lind.)
8. margaritaceum.
9. dioicum.
14. CHRYSANTHEMUM.
 1. leucanthemum.
 2. segetum.

 (Pyrethrum. Sm.)
3. parthenium.
4. inodorum.
5. maritimum.
15. MATRICARIA.
 1 chamomilla.
16. MARUTA.
 (Anthemis. Sm.)
 1. cotula.
 (fœtida. Lind.)
17. ANTHEMIS.
 1. nobilis.
 2. arvensis.
 3. maritima.
 4. tinctoria.
18. ACHILLEA.
 1. tomentosa.
 2. ptarmica.
 3. serrata.
 4. millefolium.
19. ARTEMISIA.
 1. absinthium.
 2. maritima.
 3. campestris.
 *4. cærulescens.
 5. vulgaris
 6. gallica.
20. TANACETUM.
 1. vulgare.
  α. tenuifolium.
  β. crispum.
21. DIOTIS.
 1. maritimus.

22. XANTHIUM.
 *1. strumarium.
23. BIDENS.
 1. tripartita.
 β. radiata.
 2. cernua.
 β. radiata.
24. ARCTIUM.
 (Lappa. Lind.)
 1. lappa.
 (L. glabra. Lind.)
 2? bardana.
 (L. tomentosa. Lind.)
25. ONOPORDUM.
 1. acanthium.
26. SILYBUM.
 (Carduus. Sm.)
 1. marianum.
27. CARDUUS.
 1. nutans.
 2. acanthoides.
 3. tenuiflorus.
28. SERRATULA.
 1. tinctoria.
29. SAUSSUREA.
 (Serratula. Sm.)
 1. alpina.
30. CNICUS.
 (Cirsium. D. C.)
 1. palustris.
 2. pratensis.
 3. lanceolatus.
 4. eriphorus.
 5. acaulis.
 6. arvensis.
 7. forsteri.
 8. tuberosus.
 9. heterophyllus.
31. CENTAUREA.
 1. jacea.
 2. nigra.
 β. radiata.
 3. cyanus.
 4. scabiosa.
 5. isnardi.
 *6. solstitialis.
 7. calcitrapa.

32. CARLINA.
 1. vulgaris.
33. SONCHUS.
 1. cæruleus.
 2. palustris.
 3. arvensis.
 4. oleraceus.
 α. lævis.
 β. asper.
34. LACTUCA.
 1. virosa.
 2. scariola.
 3. saligna.
35. CHONDRILLA.
 (Prenanthes. Sm.)
 1. muralis.
36. PRENANTHES.
 (Crepis. Sm.)
 1. pulchra.
 (hieracifolia. Lind.)
37. LAPSANA.
 1. communis.
 2. pusilla.
38. BARKHAUSIA.
 (Crepis. Sm.)
 1. fœtida.
39. CREPIS.
 1. tectorum.
 2. biennis.
40. LEONTODON.
 (Taraxacum. D. C.)
 1. taraxacum.
 (T. dens leonis. D. C.)
 α. officinale.
 β. angustifolium.
 2? palustre.
41. HELMINTHIA.
 (Picris. Sm.)
 1. echioides.
42. PICRIS.
 1. hieracioides.
43. HIERACIUM.
 1. alpinum.
 2. pilosella.
 3. auricula.
 β. dubium. D. C.
 4. aurantiacum.
 5. murorum.

15

β. *laciniatum.*
6. *maculatum.*
7. *sylvaticum.*
8. *pulmonarium.*
9. *lawsoni.*
10. *molle.*
11. *cerinthioides.*
12. *villosum.*
13. *halleri.*
14. sabaudum.
15. *denticulatum.*
16. *prenanthoides.*
17. umbellatum.
18. *paludosum.*
44. HYPOCHÆRIS.
1. maculata.
2. radicata.
3. glabra.
45. TRAGOPOGON.
1. pratensis.
2. *porrifolius.*
46. APARGIA.
1. *taraxaci.*
2. *alpina.*
3. autumnalis.
4. hispida.
5. *hirta.*
(*Thrincia hir:* Lind.)
47. CICHORIUM.
1. intybus.
XLV. LOBELIACEÆ.
1. LOBELIA.
1. *dortmanna.*
2. *urens.*
XLVI. CAMPANULACEÆ.
1. JASIONE.
1. montana.
2. PHYTEUMA.
1. *orbiculare.*
2. *spicatum.*
3. PRISMATOCARPUS. *D.C.*
(*Campanula.* Sm.)
1. hybridus.
4. CAMPANULA.
1. rotundifolia.
2. *patula.*
3. *rapunculus.*
4. *persicifolia.*

5. latifolia.
6. trachelium.
7. *rapunculoides.*
8. glomerata.
9. *hederacea.*
XLVII. VACCINIEÆ.
1 VACCINIUM.
1. *uliginosum.*
2. *myrtillus.*
3. *vitis-idæa.*
4. oxycoccus.
XLVIII. ERICINEÆ.
(*EMPETREÆ.* Lind.)
1? *Empetrum.*
1. nigrum.

(*PYROLEÆ.* Lind.)
2. PYROLA.
1. *uniflora.*
2. *secunda.*
3. *minor.*
4. *media.*
5. *rotundifolia.*

(*ERICEÆ.* Lind.)
3. ARBUTUS.
1. *unedo.*

(*Arctostaphylos.* Lind.)
2. *alpina.*
3. *uva ursi.*
4. ANDROMEDA.
1. *polifolia.*
5. ERICA.
1. *vagans.*
2. cinerea.
3. tetralix.
4. *ciliaris.*
6. CALLUNA.
1. vulgaris.
7. MENZIESIA.
1. *dabeoci.*
2. *cærulea.*
8. AZALEA.
1. *procumbens.*
9. LEDUM.
1. *palustre.*
XLIX. MONOTROPEÆ.
(*PYROLEÆ.* Lind.)

1. MONOTROPA.
  1. hypopitys.
L. JASMINEÆ.
  (*OLEINEÆ.* *Lind.*)
  1. LIGUSTRUM.
    1. vulgare.
  2. FRAXINUS.
    1. excelsior.
    2. *heterophyllus.*
LI. APOCYNEÆ.
  1. VINCA.
    1. major.
    2. minor.
LII. GENTIANEÆ.
  1. MENYANTHES.
    1. trifoliata.
  2. VILLARSIA.
    (*Menyanthes.* *Sm.*)
    1. nymphæoides.
  3. CHLORA.
    1. perfoliata.
  4. SWERTIA.
    1. *perennis.*
  5. GENTIANA.
    1. *pneumonanthe.*
    *2. *acaulis.*
    3. *verna.*
    4. *nivalis.*
    5. amarella.
      α. autumnalis.
      β. *vernalis.*
    6. *campestris.*
  6. ERYTHRÆA.
    1. centaureum.
    2. *latifolia.*
    3. pulchella.
    4. *littoralis.*
  7. EXACUM.
    1. *filiforme.*
LIII. POLEMONIACEÆ.
  1. POLEMONIUM.
    1. *cœruleum.*
LIV. CONVOLVULACEÆ.
  1. CONVOLVULUS.
    1. arvensis.

    (*Calystegia.* *Lind.*)
    2. sepium.
    3. *soldanella.*

2. CUSCUTA.
  1. europæa.
  2. epithymum.
LV. BORAGINEÆ.
  1. ECHIUM.
    1. vulgare.
  2. LITHOSPERMUM.
    1. arvense.
    2. officinale.
    3. *purpuro-cœruleum.*
    4. *maritimum.*
  3. PULMONARIA.
    1. *angustifolia.*
    2. *officinalis.*
  4. SYMPHYTUM.
    1. officinale.
      β. *patens.*
    2. *tuberosum.*
  5. LYCOPSIS.
    1. arvensis.
  6. ANCHUSA.
    1. *officinalis*
    2. sempervirens.
  7. BORAGO.
    1. officinalis.
  8. ASPERUGO.
    1. procumbens.
  9. MYOSOTIS.
    1. palustris.
    2. *cæspitosa.*
    3. *intermedia.*
    4. sylvatica.
    5. *alpestris.*
    6. arvensis.
    7. versicolor.
  10. CYNOGLOSSUM.
    1. officinale.
    2. *sylvaticum.*
LVI. SOLANEÆ.
  1. SOLANUM.
    1. nigrum.
    2. dulcamara.
  2. ATROPA.
    1. belladonna.
  3. DATURA.
    *1. stramonium.
  4. HYOSCYAMUS.
    1. niger.

5. VERBASCUM.
  1. thapsus.
  2? *thapsiforme.* *D. C.*
  3 lychnitis.
  4. *pulverulentum.*
  5. nigrum.
  6. *virgatum.*
  7. *blattaria.*
LVII. ANTIRRHINEÆ.
(*SCROPHULARINEÆ. Lind.*)
  1. DIGITALIS.
    1. *purpurea.*
  2. ANTIRRHINUM.
    *1. majus.
    2. *orontium.*
  3. LINARIA.
    (*Antirrhinum.* *Sm.*)
    *1. cymbalaria.
    2. elatine.
    3. spuria.
    β. peloria.
    4. minor.
    5. *repens.*
    6. vulgaris.
    (*A. linaria.* *Sm.*)
    β. peloria.
    γ. ecalcarata.
  4. SCROPHULARIA.
    1. nodosa.
    2. aquatica.
    3. *vernalis.*
    4. *scorodonia.*
  5. LIMOSELLA.
    1. aquatica.
LVIII. OROBANCHEÆ.
  1. OROBANCHE.
    1. major.
    2. elatior.
    3. minor.
    4. *rubra.*
    5. *cærulea.*
    6. ramosa.
  2. LATHRÆA.
    1. *squamaria.*
LIX. RHINANTHACEÆ.
  1. MELAMPYRUM.
(*MELAMPYRACEÆ. Lind.*)
    1. cristatum.

  2. *arvense.*
  3. pratense.
  4. *sylvaticum.*

(*SCROPHULARINEÆ. Lind.*)
  2. PEDICULARIS.
    1. palustris.
    2. sylvatica.
  3. RHINANTHUS.
    1. crista-galli.
    2? *major.*
  4. BARTSIA.
    1. *alpina.*
    2. *viscosa.*
  5. EUPHRASIA.
    1. officinalis.
    2. odontites. *D. C.*
    (*Bartsia od.* *Sm.*)
  6. SIBTHORPIA.
    1. *europæa.*
LX. VERONICEÆ.
(*SCROPHULARINEÆ. Lind.*)
  1. VERONICA.
    1. spicata.
    2? *hybrida.*
    3. *fruticulosa.*
    4. *saxatilis.*
    5. *alpina.*
    6. serpyllifolia.
    β. *humifusa.*
    7. beccabunga.
    8. anagallis.
    9. scutellata.
    10. officinalis.
    β. *allionii.*
    11. *hirsuta.*
    12. chamædrys.
    13. montana.
    14. *triphyllos.*
    15. hederifolia.
    16. agrestis.
    17. arvensis.
    18. *verna.*
LXI. LABIATÆ.
  1. LYCOPUS.
    1. europæus.
  2. SALVIA.
    1. verbenaca.

c

2. *pratensis.*
3. AJUGA.
   1. chamæpitys.
   2. *alpina.*
   3. reptans.
   4. *pyramidalis.*
4. TEUCRIUM.
   1. scorodonia.
   *2. *chamædrys.*
   3. scordium.
5. GALEOBDOLON.
   1. luteum.
6. LEONURUS.
   1. cardiaca.
7. MARRUBIUM.
   1. vulgare.
8. BALLOTA.
   1. nigra.
9. BETONICA.
   1. officinalis.
10. GALEOPSIS.
    1. tetrahit.
    β. versicolor. *D.C.*
    2. *villosa.*
    3. ladanum.
11. LAMIUM.
    1. album.
    2. *maculatum.*
    3. purpureum.
    4. *incisum.*
    5. amplexicaule.
12 NEPETA.
    1. cataria.
13. STACHYS.
    1. arvensis.
    2. *ambigua.*
    3. sylvatica.
    4. palustris.
    5. *germanica.*
14. GLECHOMA.
    1. hederacea.
15. MENTHA.
    1. *viridis.*
    α. *spicata.*
    β. *hirsutior.*
    2. piperita.
    α. officinalis.
    β. *ovata.*

γ. *hircina.*
3. sylvestris.
   α. longifolia.
   β. villosa.
   γ. *candicans.*
   δ. *nemorosa.*
4. rotundifolia.
5. *citrata.*
6. hirsuta.
   α. aquatica.
   β. palustris.
   γ. paludosa.
   δ. sativa.
7. *acutifolia.*
8. *rubra.*
9. *gentilis.*
10. gracilis.
11. arvensis.
12. agrestis.
13. pulegium.
16. THYMUS.
    1. serpyllum.
    α. majus.
    β. hirsutum.
    2. acinos.
    3. nepeta.
    4. calamintha.
17. MELITTIS.
    1. *melissophyllum.*
    2? *grandiflora.*
18. CLINOPODIUM.
    1. vulgare.
19. ORIGANUM.
    1. vulgare.
20. PRUNELLA.
    1. vulgaris.
21. SCUTELLARIA.
    1. *minor.*
    2. galericulata.
LXII. VERBENACEÆ.
1. VERBENA.
   1. officinalis.
LXIII. LENTIBULARIÆ.
1. PINCUIGULA.
   1. vulgaris.
   2. *grandiflora.*
   3. *lusitanica.*

2. UTRICULARIA.
1. vulgaris.
2. *intermedia.*
3. minor.
LXIV. PRIMULACEÆ.
1. HOTTONIA.
1. palustris.
2. LYSIMACHIA.
1. vulgaris.
2. *thyrsiflora.*
3. nemorum.
4. nummularia.
3. CENTUNCULUS.
1. minimus.
4. ANAGALLIS.
1. arvensis. *D.C.*
α. phœnicea.
β. cærulea.
2. tenella.
5. TRIENTALIS.
1. *europæa.*
6. PRIMULA.
1. veris. *Linnæus.*
α· officinalis.
β. elatior.
γ· acaulis.
(*P. grandiflora. D.C.*)
2. *farinosa.*
3? *scotica.*
7. CYCLAMEN.
*1. *hederifolium.*
8. GLAUX.
1. maritima.

9. SAMOLUS.
1. valerandi.
LXV. PLUMBAGINEÆ.
1. STATICE.
1. limonium.
β. *parva.*
2. reticulata.
3. armeria. *D.C.*
(*Armeria maritima. Lind.*)
α. pubescens.
β. *alpina.*
2. LITTORELLA.
1. lacustris.

LXVI. PLANTAGINEÆ.
1. PLANTAGO.
1. coronopus.
2. maritima.
3. lanceolata.
4. media.
5. major.
LXVII.AMARANTHACEÆ
1. AMARANTHUS.
*1. blitum.
LXVIII. CHENOPODEÆ.
1. SALICORNIA.
1. herbacea. *(Hooker.)*
β. *procumbens.*
2. *radicans. (Hooker.)*
β. *fruticosa.*
2. SALSOLA.
1. *kali.*
3. CHENOPODIUM.
1. *fruticosum.*
(*Salsola fr. Sm.*)
2. maritimum.
3. polyspermum.
4. acutifolium.
5. olidum.
(*vulvaria. D.C.*)
6. *glaucum.*
7. *botryodes.*
8. album.
9. hybridum.
10. rubrum.
11. ficifolium.
12. murale.
13. urbicum.
14. bonus-henricus.
4. ATRIPLEX.
1. portulacoides.
2. pedunculata.
3. *laciniata.*
4. patula.
5. angustifolia.
6. littoralis.
β. *erecta. D. C.*
5. BETA.
1. maritima.
LXIX. POLYGONEÆ.
1. RUMEX.
1. maritimus.

2. palustris.
3. pulcher.
4. obtusifolius.
5. acutus.
   (glomeratus. Lind.)
6. sanguineus.
   β. viridis.
   (*R. nemolapathum. Lind.*)
7. crispus.
8. hydrolapathum.
9. acetosa.
10. acetosella.
2. OXYRIA.
   1 *reniformis.*
3. POLYGONUM.
*1. fagopyrum.
2. convolvulus.
3. *viviparum.*
4. bistorta.
5. amphibium.
6. hydropiper.
7. persicaria.
8. lapathifolium.
9. minus.
10. aviculare.
LXX. THYMELEÆ.
1. DAPHNE.
   1. *mezereum.*
   2. laureola.
LXXI. SANTALACEÆ.
1. THESIUM.
   1. linophyllum.
LXXII. ELÆAGNEÆ.
1. HIPPOPHAE.
   1. *rhamnoides.*
LXXIII. ARISTOLOCHIEÆ.
   (*ARISTOLOCHIÆ. Lind.*)
1. ARISTOLOCHIA.
   1. clematitis.
2. ASARUM.
   1. *europæum.*
LXXIV. EUPHORBIACEÆ
1. BUXUS.
   1. *sempervirens.*
2. EUPHORBIA.
   1. *peplis.*
   2. helioscopia.
   3. platyphylla. *D.C.*

α. micrantha.
β. stricta.
4. *hiberna.*
5. *esula.*
6. *cyparissias.*
7. *portlandica.*
   (*segetalis.*
   β. *maritima. Lind.*)
8. *paralias.*
   (*paralia. Sm.*)
9. exigua.
10. peplus.
*11. *lathyris.*
12. *characias.*
13. amygdaloides.
3. MERCURIALIS.
   1. perennis.
   2. annuus.
LXXV. URTICEÆ.
1. PARIETARIA.
   1. officinalis.
2. URTICA.
   1. *pilulifera.*
   2. urens.
   3. dioica.
3. HUMULUS.
   1. lupulus.
LXXVI. AMENTACEÆ.
   (*ULMACEÆ. Lind.*)
1. ULMUS.
   (*Lindley.*)
   1. campestris. *Sm.*
   (α. *microphylla. D.C.*)
   2. *suberosa. Sm.*
   (*camp: β? D. C.*)
   3. *major. Sm.*
   (*camp: γ? D. C.*)
   4. *carpinifolia.*
   5. glabra. *Sm.*
   β. *glandulosa.*
   γ. *latifolia.*
   6. *stricta.*
   β. *parvifolia.*
   7. montana. *Sm.*

   (*AMENT: VERÆ. Lind.*)
2. BETULA.
   1. *nana.*

2. alba.
  β. pendula.
3. ALNUS.
  1. glutinosa.
4. SALIX.
  1. triandra.
  2. *hoffmanniana.*
  3. lanceolata.
  4. amygdalina.
  5. *pentandra.*
  6. *nigricans.*
  7. *phylicifolia.*
  8. *borreriana.*
  9. *nitens.*
  10. *davalliana.*
  11. *wulfeniana.*
  12. *tetrapla.*
  13. *bicolor.*
  14. *tenuifolia.*
  15. *malifolia.*
  16. *petiolaris.*
  17. *vitellina.*
  18. *decipiens.*
  19. fragilis.
  20. russelliana.
  21. purpurea.
  22. helix.
  23. *lambertiana.*
  24. forbiana.
  25. rubra.
  26. *croweana.*
  27. *prunifolia.*
  28. *vacciniifolia.*
  29. *venulosa.*
  30. *myrsinites.*
  31. *dicksoniana.*
  32. *carinata.*
  33. *arbuscula.*
  34. *livida.*
  35. *herbacea.*
  36. *reticulata.*
  37. *glauca.*
  38. *stuartiana.*
  39. *arenaria.*
  40. *lanata.*
  41. *argentea.*
  42. *fœtida.*
  43. repens.

44. *fusca.*
45. *prostrata.*
46. *incubacea.*
47. *doniana.*
48. *rosmarinifolia.*
49. *cinerea.*
50. aurita.
51. aquatica.
52. oleifolia.
53. *cotinifolia.*
54. *hirta.*
55. *rupestris.*
56. *andersoniana.*
57. *forsteriana.*
58. *sphacelata.*
59. caprea.
60. acuminata.
61. viminalis.
62. *smithiana.*
63. *stipularis.*
64. alba.
5. POPULUS
  1. alba.
  2. canescens.
  3. tremula.
  4. nigra.
    β. *viridis.*

(CUPULIFERÆ.   Lind.)
6. FAGUS.
  1. sylvatica.
7. CASTANEA.
  1. *vulgaris.*
    (Fagus cast :   Sm.)
8. QUERCUS.
  1. robur.
  2. *sessiliflora.*
9. CORYLUS.
  1. avellana.
10. CARPINUS.
  1. betulus.

LXXVII. MYRICEÆ.
  1. MYRICA.
    1. gale.

LXXVIII. CONIFERÆ.
  1. TAXUS.
    1. baccata.

2. JUNIPERUS.
  1. communis.
  2? nana.
3. PINUS.
  1. sylvestris.

MONOCOTYLEDONES
OR
ENDOGENÆ.

LXXIX. HYDROCHARIDEÆ.
  1. STRATIOTES.
    1. aloides.
  2. HYDROCHARIS.
    1. morsus-ranæ.
LXXX. ALISMACEÆ.
  (BUTOMEÆ. Lind.)
  1. BUTOMUS.
    1. umbellatus.

(ALISM: VERÆ. Lind.)
  2. ALISMA.
    1. ranunculoides.
    β. repens.
    2. natans.
    3. plantago.
    4. damasonium.
  3. SAGITTARIA.
    1. sagittifolia.

(JUNCAGINEÆ. Lind.)
  4. SCHEUCHZERIA.
    1. palustris.
  5. TRIGLOCHIN.
    1. maritimum.
    2. palustre.
LXXXI. POTAMEÆ.
  (FLUVIALES. Lind.)
  1. POTAMOGETON.
    1. natans.
    β. fluitans. D. C.
    2. proteum. Lind.
    α. lucens.
    β. heterophyllum.
    3. fluitans. Sm.
    (obscurum. D. C.)
    (rufescens. Lind.)

4. lanceolatum.
5. perfoliatum.
6. crispum.
7. densum.
8. compressum.
β? cupisdatum. D. C.
(P. zosterifolium. Lind.)
9. gramineum.
(obtusifolium. Lind.)
10. pusillum.
11. pectinatum.
2. RUPPIA.
  1. maritima.
3. ZANNICHELLIA.
  1. palustris.
4. ZOSTERA.
  1. marina.
LXXXII. ORCHIDEÆ.
  1. ORCHIS.
    1. morio.
    2. mascula.
    3. ustulata.
    4. militaris.
    5? fusca. D.C.
    6. tephrosanthos.
    (simia. D. C.)
    7. pyramidalis.
    (Anacamptis pyr: Lind.)
    8. hircina:
    9. latifolia.
    10. maculata.
  2. GYMNADENIA.
    (Orchis. D.C. Sm.)
    1. conopsea.
  3. ACERAS.
    (Ophrys. D.C.)
    1. anthropophora.
  4. HERMINIUM.
    (Ophrys. D.C.)
    1. monorchis.
  5. HABENARIA.
    (Orchis. D. C.)
    (Platanthera. Lind.)
    1. viridis.
    2. albida.
    3. bifolia.
  6. OPHRYS.
    1. apifera.

2. aranifera.
3. *arachnites.*
4. *fucifera.*
5. muscifera.
7. GOODYERA.
 (*Neattia.* *D.C.*)
 1. *repens.*
8. NEOTTIA.
 (*Spiranthes.* *Lind.*)
 1. spiralis.
 (*S. autumnalis.* *Lind.*)
 2. *gemmipara.*
9. LISTERA.
 1. ovata.
 2. *cordata.*
 3. nidus-avis.
 (*Neottia nid:* *Lind.*)
 (*Epipactis nid:* *Sm.*)
10. EPIPACTIS.
 1. latifolia.
 2. *purpurata.*
 3. palustris.
 4. grandiflora.
 (*pallens.* *D. C.*)
 5. *ensifolia.*
 6. *rubra.*
11. MALAXIS.
 1. loeselii.
 (*Liparis loes.* *Lind.*)
 2. paludosa.
12. CORALLORRHIZA.
 1. *innata.*
 (*halleri.* *D.C.*)
13. CYPRIPEDIUM.
 1. *calceolus.*
LXXXIII. IRIDEÆ.
 1. IRIS.
 1. pseudacorus.
 2. fœtidissima.
 2. TRICHONEMA.
 1. *bulbocodium*
 3. CROCUS.
 1. *vernus.*
 2. *sativus.*
 3. *reticulatus.*
 4. *nudiflorus.*
LXXXIV. AMARYLLIDEÆ
 1. NARCISSUS.

1. pseudo-narcissus.
2. *poeticus.*
3. *biflorus.*
2. LEUCOJUM.
 1. *æstivum.*
3. GALANTHUS.
 1. *nivalis.*
LXXXV. ASPARAGEÆ.
 (*ASPHODELEÆ.* *Lind.*)
 1. ASPARAGUS.
 1. *officinalis.*

 (*SMILACEÆ.* *Lind.*)
 2. PARIS.
 1. quadrifolia.
 3. CONVALLARIA.
 1. *verticillata.*
 2. *polygonatum.*
 3. *multiflora.*
 4. majalis.
 4. RUSCUS.
 1. aculeatus.
 5. TAMUS.
 1. communis.
LXXXVI. LILIACEÆ.
 1. TULIPA.
 1. *sylvestris.*
 2. FRITILLARIA.
 1. meleagris.

 (*ASPHODELEÆ.* *Lind.*)
 3. ANTHERICUM.
 (*Phalanguim.* *D. C.*)
 1. *serotinum.*
 4. SCILLA.
 1. *autumnalis.*
 2. bifolia.
 3. verna.
 4. nutans.
 (*Hyacinthus non-scriptus.Lind.*)
 5. MUSCARI.
 *1. *racemosum.*
 6. GAGEA.
 (*Ornithogalum.* *Sm.*)
 1. *lutea.*
 7. ORNITHOGALUM.
 1. umbellatum.
 *2. *nutans.*
 3. pyrenaicum.

8. ALLIUM.
*1. *ampeloprasum.*
2. vineale.
3. *arenarium.*
4. oleraceum.
5. *schœnoprasum.*
6. *carinatum.*
7. ursinum.

LXXXVII.COLCHICACEÆ
(*MELANTHACEÆ.* Lind.)
1. COLCHICUM.
1. autumnale.
2. TOFIELDIA.
1. *palustris.*

LXXXVIII. JUNCEÆ.
1. NARTHECIUM.
(*Abama. D. C.*)
1. ossifragum.
2. JUNCUS.
1. *filiformis.*
2. *arcticus.*
3. communis. *D.C.*
α. conglomeratus.
β. effusus.
4. glaucus.
5. *acutus.*
6. *maritimus.*
7. squarrosus.
8. compressus.
9. bufonius.
10. *gesneri.*
11. *trifidus.*
12. *cœnosus.*
13. uliginosus.
β. viviparus.
14. *capitatus.*
15. *subverticillatus.*
16. lampocarpus.
17. acutiflorus.
18. obtusiflorus.
19. *polycephalus.*
20. *triglumis.*
21. *biglumis.*
22. *castaneus.*
3. LUZULA.
(*Luciola. Sm.*)
1. pilosa.

2. *forsteri.*
3. *sylvatica.*
4. *arcuata.*
5. *spicata.*
6. campestris.
β. congesta. *D. C.*

LXXXIX. RESTIACEÆ.
1. ERIOCAULON.
1. *septangulare.*
XC. AROIDEÆ.
1. ARUM.
1. maculatum.
2. ACORUS.
1. calamus.
XCI. TYPHACEÆ.
1. TYPHA.
1. latifolia.
2. *minor.*
3. angustifolia.
2. SPARGANIUM.
1. natans.
2. simplex.
3. ramosum.
XCII. CYPERACEÆ.
1. CYPERUS.
1. *longus.*
2. *fuscus.*
2. SCHÆNUS.
1. nigricans.
3. RHYNCHOSPORA.
(*Schœnus. D. C.*)
1. alba.
2. *fusca.*
4. CLADIUM.
(*Schœnus. D. C.*)
1. mariscus.
5. ELEOCHARIS.
(*Heleocharis. Lind.*)
(*Scirpus. D. C.*)
1. palustris.
2. multicaulis.
3. acicularis.
_____

(*Scirpus. Sm. D.C.*)
4. cæspitosa.
5. pauciflora.
6. ISOLEPIS.
(*Scirpus. D. C.*)

1. *fluitans.*
    (*Heliogiton fl.* *Lind.*)
2. setacea.
3. *holoschænus.*
(*Holoschænus vulgaris.* *Lind.*)
7. BLYSMUS.
    (*Schœnus.* *D. C.*)
    1. compressus.
(*Scirpus caricinus.* *Sm.*)
8. SCIRPUS.
    1. lacustris.
    2. *glaucus.*
    3. *rufus.*
        (*Schœnus ruf:* *Lind.*)
    4. *triquetur.*
    5. *carinatus.*
    6. maritimus.
    7. *sylvaticus.*
9. ERIOPHORUM.
    1. *polystachion.*
    2. pubescens.
    3. angustifolium.
    4. *gracile.*
    5. *vaginatum.*
    6. *capitatum.*
    7. *alpinum.*
10. KOBRESIA.
    1. *caricina.*
        (*scirpina.* *D.C.*)
11. CAREX.
    1. dioica.
        β. capitata.
    2. *davalliana.*
    3. pulicaris.
    4. *pauciflora.*
    5. stellulata
    6. curta.
    7. *elongata.*
    8. ovalis.
    9. *tenella.*
    10. remota.
    11. axillaris.
    12. *incurva.*
        (*juncifolia.* *D.C.*)
    13. *arenaria.*
    14. intermedia.
        (*disticha.* *D.C.*)
    15. *divisa.*

16. muricata.
17? divulsa. *D.C.*
18. vulpina.
19. teretiuscula.
20. paniculata.
21. *digitata.*
22. *clandestina.*
23. pendula.
    (*maxima.* *D.C.*)
24. strigosa.
25. sylvatica.
    (*patula.* *D.C.*)
26. *depauperata.*
27. *melichofera.*
28. *speirostachya.*
29. *phæostachya.*
30. *capillaris.*
31. *rariflora.*
32. pseudo-cyperus.
33. *limosa.*
34. *ustulata.*
35. atrata.
36. *pulla.*
37. pallescens.
38. flava.
39? oederi.
    (*flava.* β. *D. C.*)
40. fulva.
41. *extensa.*
42. distans.
43? binervis *D.C.*
44. præcox.
45. pilulifera.
46. *tomentosa.*
47. panicea.
48. recurva.
    (*glauca.* *D.C.*)
    β. micheliana.
49? *rigida.*
50. cæspitosa.
51. stricta.
52. acuta.
53. paludosa.
54. riparia.
55. *lavigata.*
56. vesicaria.
57. ampullacea
58. hirta.

D

59. *secalina.*
60. *stictocarpa.*
61. *angustifolia.*
62. filiformis.
XCIII. GRAMINEÆ.
1. LAGURUS.
  1. *ovatus.*
2. CYNODON.
  1. *dactylon.*
3. DIGITARIA.
  1. *sanguinalis.*
4. CALAMAGROSTIS.
    (*Arundo.* *Sm.*)
  1. epigejos.
  2. lanceolata.
    (*A. calamag.* *Sm.*)
  3. *stricta.*

    (*Ammophila.* *Lind.*)
4. *arenaria.*
5. AGROSTIS.
  1. alba.
    β. stolonifera.
  2. vulgaris.

    (*Trichodium.* *Lind.*)
  3. canina.
  4. *setacea.*

    (*Anemagrostis.* *Lind.*)
  5. spica-venti.
6. MILIUM.
  1. *lendigerum.*
  2. effusum.
    (*Gastridium ef.* *Lind.*)
7. STIPA.
  *1. *pennata.*
8. PANICUM.
    (*Echinochloa.* *Lind.*)
  1. *crus-galli.*

    (*Setaria.* *Lind.*)
  2. viride.
  3. *verticillatum.*
9. DIGRAPHIS.
  (*Phalaris.* *Sm: D.C.*)
  1. arundinacea.
10. PHALARIS.
  *1. *canariensis.*

11. PHLEUM.
  1. pratense.
  β. nodosum.
  2. asperum.
  3. *alpinum.*

    (*Phalaris.* *D. C.*)
  4. *michelii.*
    (*alpinum.* *D. C.*)
  5. boehmeri.
    (*phleoides.* *D. C.*)
  9. arenaria.
    (*Achnodon ar:* *Lind.*)
12. POLYPOGON.
  1. *monspeliensis.*
  2. *littoralis.*
13. ALOPECURUS.
  1. pratensis.
  2. *alpinus.*
  3. agrestis
  4. *bulbosus.*
  5. geniculatus.
    β? fulvus.
14. ANTHOXANTHUM.
  1. adoratum.
15. HIEROCHLOE.
  1. *borealis.*
16. MOLINIA.
    (*Melica.* *Sm.*)
  1. cærulea.
  2. *depauperata.*
17. MELICA.
  1. uniflora.
  2. *nutans.*
18. AIRA.
  1. *alpina.*
  2. flexuosa.
  3 caryophyllea.
  4. præcox.

    (*Deschampsia.* *Lind.*)
  5. cæspitosa.

    (*Corynephorus.* *Lind.*)
  6. canescens.
19. HOLCUS.
    (*Avena.* *D. C.*)
  1. lanatus.
  2. mollis.

20. ARRHENATHERUM.
(*Avena. D.C.*)
1. elatius.
(*Holcus avenac.* Sm.)
2. *bulbosum.*
21. AVENA.
1. fatua.
2. *strigosa.*
3. *alpina.*
4. pratensis.

(*Trisetum. Lind.*)
5. pubescens.
6. flavescens.
22. TRIODIA.
(*Danthonia. D.C.*)
1. decumbens.
23. BROMUS.
1. secalinus.
2. velutinus.
3. mollis.
*4. *squarrosus.*
5. erectus.
6. *arvensis.*
7. racemosus.
β. pratensis.
8. asper.
9. sterilis.
10. *diandrus.*
24. FESTUCA.
1. ovina.
2? *vivipara.*
3. duriuscula.
4? *rubra. D.C.*
5. gigantea.
β. *triflora.*

(*Schenodorus. Lind.*)
6. pratensis.
7. elatior.
8. *calamaria.*
(*sylvatica. D.C.*)
β. *decidua.*
9. loliacea.

(*Vulpia. Lind.*)
10. *uniglumis.*
11. myurus.
12. *bromoides.*

25. ARUNDO.
1. phragmitês.
26. DACTYLIS.
1. glomerata.
27. KŒLERIA.
(*Aira. Sm.*)
(*Airochloa. Lind.*)
1. cristata.
28. GLYCERIA.
(*Poa. D.C.*)
1. fluitans.

(*Hydrochloa. Lind.*)
2. aquatica.

(*Schlerochloa. Lind.*)
3 maritima.
β. distans. D.C.
(*Poa dis: Lind.*)
4. *procumbens.*
5. rigida.
29. POA.
1. compressa.
2. *bulbosa.*
3. trivialis.
4. pratensis.
β. *subcærulea.*
γ. angustifolia.
5. nemoralis.
β. *glauca. D.C.*
γ. *cæsia.*
6. annua.
7. *laxa.*
8. *alpina.*
30. CATABROSA.
(*Aira. Sm.*)
1. aquatica.
(*Poa airoides. D. C.*)
31. BRIZA.
1. media.
2. *minor.*
32. CYNOSURUS.
1. cristatus.
2. *echinatus.*
33. SESLERIA.
1. *cærulea.*
34. SPARTINA.
(*Trachynotia. D.C.*)
1. *stricta.*

35. KNAPPIA.
 1. *agrostidea.*
(*Chamagrostis minima. D. C.*)
36. NARDUS.
 1. stricta.
37. ROTTBOLLIA.
 (*Ophiurus. Lind.*)
 1. incurvata.
38. TRITICUM.
 1. *cristatum.*
 2. caninum.
 3. repens.
 4. *junceum.*
39. BRACHYPODIUM.
 (*Triticum. D.C.*)
 (*Festuca. Sm.*)
 1. pinnatum.
 2. sylvaticum.
40. CATOPODIUM.
 (*Triticum. Sm.*)
 1. loliaceum.
41. LOLIUM.
 1. perenne.
 2. temulentum.
 3? arvense.
42. ELYMUS.
 1. *arenarius.*
 2. *geniculatus.*
 3. *europæus.*
43. HORDEUM.
 1. murinum.
 2. pratense.
 (*secalinum. D. C.*)
 2. maritimum.
XCIV? LEMNACEÆ.
 (*PISTIACEÆ. Lind.*)
 1. LEMNA.
 1. trisulca.
 2. minor.
 3. gibba.
 4. polyrhiza.

ACOTYLEDONES
OR
CRYPTOGAMÆ.

XCV? CHARACEÆ.
 1. CHARA.

1. vulgaris.
 2. hispida.
 3? *aspera. (Greville.)*
 4. *flexilis.*
 5. *translucens.*
 6. nidifica.
 7. gracilis.
XCVI. EQUISETACEÆ.
 (*FILICES. Sm.*)
 1. EQUISETUM.
 1. arvense.
 2. fluviatile.
 3. sylvaticum.
 4. palustre.
 β. polystachyon.
 γ. *nudum.*
 5. limosum.
 6. hyemale.
 7 *variegatum.*
 (*multiforme α. D.C.*)
XCVII. FILICES.
 (*LYCOPODIACEÆ.*
Trib. *LYCOPODIEÆ. D.C.*)
 1. LYCOPODIUM.
 1. *alpinum.*
 2. *annotinum.*
 3. *clavatum.*
 4. *selago.*
 5. inundatum.
 6. *selaginoides.*

(*FILICES VERÆ. D.C.*)
 2. BOTRYCHIUM.
 1. lunaria.
 β. *ramosa.*
 γ. *dissecta.*
 3. OPHIOGLOSSUM.
 1. vulgatum.
 4. OSMUNDA.
 1. regalis.
 5 POLYPODIUM.
 1. vulgare.
 β. *serratum.*
 γ. *proliferum.*
 δ. *cambricum.*
 2. *phægopteris.*
 3. *dryopteris.*
 4. *calcareum.*

6. WOODSIA
1. *hyperborea.*
2 *ilvensis.*
7. BLECHNUM.
1. *boreale.*
8. ASPLENIUM.
1. *lanceolatum.*
2. adiantum nigrum.
   β. *longifolium.*
3. ruta-muraria.
4. *alternifolium.*
5. *marinum.*
   β. *trapeziforme.*
6. *viride.*
   β. *ramosum.*
7. trichomanes.
8. *fontanum.*
9. *septentrionale.*
9. SCOLOPENDRIUM.
1. vulgare
   β. *multifidum.*
2. *ceterach.*
   (*Ceterach officin :* D.C.)
10. PTERIS.
1. *crispa.*
2. aquilina.
   β. *maritima.*
11. ADIANTUM.
1. *capillus-veneris.*
12. ASPIDIUM.
   (*Polystichum.* D.C.)
1. oreopteris.
   β. *pumila.*
2. *thelypteris.*
3. *cristatum.*
   (*callipteris.* D.C.)
4. *dumetorum.*
5. dilatatum.
6? *spinulosum.*
7. filix-mas.
8. lonchitis.
9. aculeatum.
   (*A. lonch. var ?* D.C.)
   β. *affinis.*
10. *lobatum.*
11. *angulare.*

(*Athyrium.* D. C.)
12. filix-fœmina.
13. *irriguum.*
13. CYSTEA.
   (*Aspidium.* D.C.)
1. *fragilis.*
2. *dentata.*
3. *angustata.*
4. *regia.*
14. TRICHOMANES.
1. *brevisetum.*
15. HYMENOPHYLLUM.
1. *tunbridgense.*

(*MARSILIACEÆ.* D. C.)
16. PILULARIA.
1. *globulifera.*

(*LYCOPODIACEÆ.*)
(*Trib. ISOETIDEÆ.* D. C.)
17. ISOETES.
1. *lacustris.*
XCVIII. MUSCI.
(*From Muscologia Britannica.*)
1. ANDRÆA.
1. *alpina.*
2. *rupestris.*
3. *rothii.*
4. *nivalis.*
2. PHASCUM.
1. serratum.
2. *crassinervium.*(*Greville.*)
3. alternifolium.
4. *crispum.*
5. subulatum.
6. *axillare.*
7. *patens.*
   β. *recurvifolium.*
8. *muticum.*
   β. *minus*
9. cuspidatum
   α. apiculatum.
   β. *piliferum.*
10. bryoides.
11. rectum.
12. curvicollum.
3. SPHAGNUM.
1. obtusifolium.

α. vulgaris.
β. minus.
γ. fluitans.
2. squarrosum.
3. acutifolium.
4. cuspidatum.
4. GYMNOSTOMUM.
1. lapponicum.
2. æstivum.
3. viridissimum.
4. curvirostrum.
5. rupestre.
6. griffithianum.
7. ovatum.
α. vulgare.
β. gracile.
8. truncatulum.
β. intermedium.
9. heimii.
10. conicum.
β. minutulum.
11. fasciculare.
12. pyriforme.
13. tenue.
14. donianum.
15. microstomum.
5. ANICTANGIUM.
1 ciliatum.
2. imberbe.
6. SCHISTOSTEGA.
1. pennata.
7. DIPHYSCIUM.
1. foliosum.
8. TETRAPHIS.
1. pellucida.
2. browniana.
9. SPLACHNUM.
1. sphæricum.
2. tenue.
3. mnioides.
α. minus.
β. majus.
4. angustatum.
5. ampullaceum.
6. vasculosum.
7. frœlichianum.
10. CONOSTOMUM.

1. boreale.
11. POTYTRICHUM.
1. undulatum.
2. hercynicum.
3. piliferum.
4. juniperinum.
5. septentrionale.
6. commune.
α. yuccæfolium.
β. attenuatum.
7. alpinum.
8. urnigerum.
9. aloides.
α. major.
β. dicksoni.
10. nanum.
12. CINCLIDOTUS.
1. fontinaloides.
13. TORTULA.
1. enervis.
2. brevirostris.
3. rigida.
4. convoluta.
5. revoluta.
6. muralis.
α. pilifera.
β. mutica.
7. ruralis.
α. vulgaris.
8. subulata.
α. acuminata.
β. obtusa.
9. unguiculata.
10. stellata.
11. cuneifolia.
12. tortuosa.
13. fallax.
α. imberbis.
β. linoides.
γ. brevicaulis.
14. gracilis.
β. viridis.
14. ENCALYPTA.
1. streptocarpa.
2. vulgaris.
3. ciliata.
α. concolor.

β. *pilifera.*
4. *rhaptocarpa.*
15. GRIMMIA.
1. apocarpa.
  α. nigro-viridis.
  β. *stricta.*
2. *maritima.*
3. *saxicola.*
4. pulvinata.
5. *trichophylla.*
6. *spiralis.*
7. *torquata.*
8. *leucophæa.*
9. *ovata.*
10. *doniana.*
11. *unicolor.*
16. PTEROGONIUM.
1. *smithii.*
2. *gracile.*
3. *filiforme.*
17. WEISSIA.
1. *splachnoides.*
2. *templetoni.*
3. *nuda.*
4. *nigrita.*
5. *starkeana.*
6. *affinis.*
7. lanceolata.
8. *latifolia.*
9. *striata.*
  α. *minor.*
  β. *major.*
10. *trichodes.*
11. *cirrata.*
12. *tenuirostris.*
13. *curvirostra.*
14. *crispula.*
15. controversa.
16. calcarea.
17. *recurvata.*
18. *pusilla.*
19. *verticillata.*
20. *acuta.*
18. DICRANUM.
1. bryoides.
  α. viridulum.
  β. *osmundioides.*
  γ. *tamarindifolium.*

2. adiantoides.
3. taxifolium.
4. *glaucum.*
5. *latifolium.*
6. *longifolium.*
7. *cerviculatum.*
8. *flexuosum.*
  β. *nigro-viride.*
9. *virens.*
10. *schreberianum.*
11. *strumiferum.*
12. *polycarpum.*
13. *falcatum.*
14. *starhii.*
15. *flavescens.*
16. *squarrosum.*
17. *pellucidum.*
18. *spurium.*
19. *crispum.*
20. *scottianum.*
21. *undulatum.*
22. scoparium.
  α. majus.
  β. *fuscescens.*
23. varium.
  α. viride.
  β. *rufescens.*
  γ. *luridum.*
24. heteromallum.
25. *subulatum.*
26. *fulvellum.*
19. TRICHOSTOMUM.
1. *patens.*
  α. *majus.*
  β. *piliferum.*
2. *lanuginosum.*
3. canescens.
4. *heterostichum.*
5. *microcarpon.*
6. *aciculare.*
7. *fasciculare.*
8. *polyphyllum.*
9. *ellipticum.*
20. GLYPHOMITRION.
1. *daviesii.*
21. LEUCODON.
1. sciuroides.
22. DIDYMODON.

1. purpureum.
2. inclinatum.
3. nervosum.
4. flexifolium.
5. glaucescens.
6. bruntoni.
7. rigidulum.
8. trifarium.
9. capillaceum.
10. heteromallum.
23. FUNARIA.
1. hygrometrica.
2. muhlenbergii.
3. hibernica.
24. ZYGODON.
1. conoideum.
25. ORTHOTRICHUM.
1. cupulatum.
2. anomalum.
3. drummondii.
4. affine.
α. majus.
β. pumilum
5. rupincola.
6. diaphanum.
7. rivulare.
8. striatum.
9. lyellii.
10. speciosum.
11. hutchinsiæ.
12. ludwigii.
13. crispum.
14. pulchellum.
26. NECKERA.
1. pumila.
2. pennata.
3. crispa.
27. ANOMODON.
1. curtipendulum.
2. viticulosum.
28. DALTONIA.
1. splachnoides.
2. heteromalla.
29. FONTINALIS.
1. antipyretica.
2. squamosa.
3. capillacea.

30. BUXBAUMIA.
1. aphylla.
31. BARTRAMIA.
1. pomiformis.
α. minor.
β. major.
2. ithyphylla.
3. gracilis.
4. fontana.
α. major.
β. marchica.
5. halleriana.
6. arcuata.
32. HOOKERIA.
1. lucens.
2. læte-virens.
33. HYPNUM.
1. trichomanoides.
2. complanatum.
3. riparium.
4. undulatum.
5. denticulatum.
α. angustifolium.
β. obtusifolium.
6. medium.
7. tenellum.
8. serpens.
9. populeum.
10. reflexum.
11. molle.
12. schreberi.
13. moniliforme.
14. catenulatum.
15. stramineum.
16. trifarium.
17. murale.
18. purum.
19. plumosum.
20. pulchellum.
21. rufescens.
22. polyanthos.
23. sericeum.
24. salebrosum.
25. lutescens.
26. nitens.
27. albicans.
28. alopecurum.

29. dendroides.
30. *curvatum.*
31. myosuroides.
32 splendens.
33. proliferum.
34. prælongum.
35. *flagellare.*
36. abietinum.
37. *blandovii.*
38. *piliferum.*
39. *blandum.*
40. rutabulum.
41. velutinum.
42. ruscifolium.
43. striatum.
44. *confertum.*
45. cuspidatum.
46. *cordifolium.*
47. *polymorphum.*
48. stellatum.
   α. majus.
   β. minus.
49. *halleri.*
50. *dimorphum.*
51. loreum.
52. triquetrum.
53. *brevirostre.*
54. squarrosum.
55. filicinum.
56. *atro-virens.*
57. *palustre.*
58. fluitans.
59. aduncum.
   α. revolvens.
   β. rugosum.
60. *uncinatum.*
61. *rugulosum.*
62. *commutatum.*
63. scorpioides.
64. *silesianum.*
65. cupressiforme.
   α. vulgare.
   β. compressum.
   γ. tenue.
66. *crista-castrensis.*
67. molluscum.

34. TIMMIA.
  1. *megapolitana.*
    α. polytrichioides.
    β. austriaca.
35. BRYUM.
  1. androgynum.
  2. *palustre.*
  3. *trichodes.*
  4. *triquetrum.*
  5. *dealbatum.*
  6. pyriforme.
  7. julaceum.
  8. *crudum.*
  9. carneum.
  10. argenteum.
  11. *zierii.*
  12. roseum.
  13. capillare.
  14. cæspiticium.
    α. major.
    β. minor.
  15. *turbinatum.*
  16. *nutans.*
  17. *elongatum.*
  18. *alpinum.*
  19. ventricosum.
  20. *demissum.*
  21. punctatum.
  22. *tozeri. (Greville.)*
  23. ligulatum.
  24. rostratum.
  25. *marginatum.*
  26. hornum.
  27. cuspidatum.

XCIX. HEPATICÆ.
  1. RICCIA.
    1. crystallina.
      α. glauca.
      β. obtusa.
    2? *fluitans.*
    3. natans.
    4. *spuria.*
  2. SPHÆROCARPUS.
    1. terrestris.
  3. ANTHOCEROS.
    1. *punctatus.*

E

4. TARGIONIA.
  1. *hypophylla.*
5. MARCHANTIA.
  1. polymorpha.
  2. conica.
  3. hemisphærica.
6. JUNGERMANNIA.
  1. *trichophylla.*
  2. *setacea.*
  3. *julacea.*
  4. *laxifolia.*
  5. *juniperina.*
  6. *hookeri.*
  7. asplenioides.
  8. *spinulosa.*
  9. *decipiens.*
  10. *doniana.*
  11. *pumila.*
  12. *lanceolata.*
  13. *cordifolia.*
  14. sphagni.
  15. *crenulata.*
  16. *sphærocarpa.*
  17. *hyalina.*
  18. *compressa.*
  19. *emarginata.*
  20 *concinnata.*
  21. *orcadensis.*
  22. *inflata.*
  23. *excisa.*
  24. ventricosa.
  25. *turneri.*
  26. bicuspidata.
  27. *byssacea.*
  28. *connivens.*
  29. *curvifolia.*
  30. *capitata.*
  31. *incisa.*
  32. pusilla.
  33. *setiformis.*
  34. *nemorosa.*
  35. *planifolia.*
  36. *umbrosa.*
  37. *undulata.*
  38. *resupinata.*
  39. *albicans.*
  40. *obtusifolia.*
  41. *dicksoni.*

  42. *minuta.*
  43. *exsecta.*
  44. *cochleariformis.*
  45. complanata.
  46. *anomala.*
  47. *taylori.*
  48. *scalaris.*
  49. *polyanthos.*
  50. *cuneifolia.*
  51. viticulosa.
  52. *trichomanis.*
  53. bidentata.
  54. *heterophylla.*
  55. *stipulacea.*
  56. *francisci.*
  57. *barbata.*
  58. *albescens.*
  59. *reptans.*
  60. *trilobata.*
  61. platyphylla.
  62. *lævigata.*
  63. *ciliaris.*
  64. *woodsii.*
  65. *tomentella.*
  66. *mackaii.*
  67. *serpyllifolia.*
  68. *hamatifolia.*
  69. *minutissima.*
  70. *calyptrifolia.*
  71. *hutchinsiæ.*
  72. dilatata.
  73. tamarisci.
  74. pinguis.
  75. *multifida.*
  76. *blasia.*
  77. epiphylla.
  78. furcata.
  79. *pubescens.*
  80. *lyellii.*
  81. *hibernica.*

Enumeration of the Plants mentioned in the Catalogue.

		Great Britain				Cambridgeshire			
		Orders	Genera	Species	Varieties	Orders	Genera	Species	Varieties
Dicotyledones	Indigenous	77	378	1099	1207	71	292	643	678
	Naturalized	1	17	45	47	1	11	23	24
Monocotyledones	Indigenous	16	105	351	371	15	79	200	213
	Naturalized	—	3	6	—	—	—	—	—
Total (Phanerogamæ)		94	503	1501	1625	87	382	866	915
Part of Acotyledones		5	60	447	499	5	37	145	150

ALPHABETICAL INDEX OF THE GENERA.

The Roman numerals refer to the Order, and the Arabic to the place of the Genus under it. The *italics* are used for the synonyms.

The Orders have been omitted to save the space, as each may readily be referred to by turning to any Genus belonging to it.

40

HODSON, PRINTER, CAMBRIDGE.

SKETCH

OF

A COURSE OF LECTURES

ON

B O T A N Y.

FOR 1833.

By Rev. J. S. HENSLOW, M.A.

PROFESSOR OF BOTANY IN THE UNIVERSITY OF
CAMBRIDGE.

I. DEMONSTRATIVE.—*On Tuesdays and Thursdays.*

INTRODUCTION.—External characters of the several
organs.

Glossology.

1. Conservative.
   (*a.*) Ascending: *stem; branches; leaves.*
   (*b*) Descending: *root.*
2. Reproductive.
   (*a.*) Flower: *floral whorls.*
   (*b.*) Fruit: *pericarp and seeds.*

Organography.

Elementary or component textures: *utricles; fibre.*
Elementary organs : —
1. Cellular tissue : *cells.*
2. Woody fibre: *closters.*
3. Vascular tissue: *ducts; trachea.*
Intercellular passages—Receptacles—Lacunæ.

Compound organs :—

1. Investing or cuticular.
   (a.) cuticle.
   (b.) stomata.
   (c.) hairs, bristles, scales, prickles.
   (d.) glands.
2. Complex.

N.B. As the first few minutes of *every* Lecture will be occupied with the demonstration of some British plant, the various forms of the complex organs will be best explained at these times.

Specimens in the following Orders will most probably be among those from which the demonstrations will be given during the present Course :—

Primulaceæ, Boragineæ, Violaceæ, Cruciferæ, *Labiatæ, Salicineæ,* Ranunculaceæ, *Rosaceæ, Aroideæ,* Compositæ, *Acerineæ, Plantagineæ,* Umbelliferæ, *Rubiaceæ, Euphorbiaceæ, Resedaceæ, Liliaceæ, Geraniaceæ,* Leguminosæ, *Polygaleæ, Halorageæ, Callitrichineæ, Cyperaceæ,* Orchideæ, *Junceæ, Gramineæ,* Filices, Musci.

Phytography.

(a.) Conservative or fundamental.
   (α.) Root : *caudex, fibrils.*
   (β.) Stem (aerial) : *branches, runners, thorns.*
   —— (subterranean) : *rhizoma, tuber, bulb.*
   (γ.) Leaf (simple and compound) : *vernation,*
   —— *petiole and limb—leaflets—stipules.*
   —— *sheath—pitchers, &c.*
   —— nervation : *angulinerved, curvinerved.*
   (δ.) Appendages : *tendrils, claws, &c.*

(b.) Reproductive.
   (α.) Inflorescence : *axillary, terminal.*
   —————— *peduncle, pedicel, involucrum, bractea.*
   —————— *Raceme, spike, catkin, cone, spadix.*
   —————— *Umbel (simple and compound).*
   —————— *Head, corymb, cyme, thyrsus.*

(β.) Flower: *perianth—æstivation*.

───── : *receptacle (torus)*.

Hypogynous, perigynous, epigynous.

(1.) Calyx: *sepals*.

(2.) Corolla: *petals (claws and limb)*.

(3.) Stamens: *filament, anther, pollen*.

(4.) Pistil: *carpels, ovarium or germen, ovules, style, stigma*.

Disk—nectary.

(γ.) Fruit.

* Pericarp: *valves, paries, placenta*.

† Apocarpous fruits: *follicle, pod (legumen), nut, drupe, &c*.

†† Syncarpous fruits: *capsule, pyxidium, siliqua, samara, &c. &c*.

** Seed: *integuments (arillus, testa), hilum, albumen*.

───── : *embryo (radicle, plumule, cotyledons)*

(c.) Herb: *duration, habit, habitat, use, &c. &c*.

─────◆─────

Taxonomy.

After the demonstration of each specimen, its Genus and Species will be ascertained, by referring it to its Class and Order in the Artificial System of Linnæus.

When the name of the plant is discovered, its right position must be sought for in the Natural System. But after some little progress has been made, this may be ascertained, without at all referring to the Linnæan System, by merely consulting some artificial analysis of the Natural Orders, like the one given by Professor Lindley in his Introduction to the Natural System.

4

Artificial System of Linnœus.

CLASSES.		ORDERS.
1. Monandria	⎫	Monogynia
2. Diandria		Digynia
3. Triandria		Trigynia
4. Tetrandia		Tetragynia
5. Pentandria		Pentagynia
6. Hexandria		Hexagynia
7. Heptandria	⎬	Heptagynia
8. Octandria		Octogynia
9. Enneandria		Enneagynia
10. Decandria		Decagynia
11. Dodecandria		Dodecagynia
12. Icosandria		Polygynia
13. Polyandria	⎭	
14. Didynamia	⎰⎱	Gymnospermia Angiospermia
15. Tetradynamia	⎰⎱	Siliculosa Siliquosa
16. Monadelphia 17. Diadelphia 18. Polyadelphia	⎰⎱	Triandria, &c. *as in the Classes*
19 Syngenesia	⎰⎱	Polygamia, æqualis ———— superflua ———— frustranea ———— necessaria ———— segregata Monogamia
20. Gynandria 21. Monœcia 22. Diœcia	⎰⎱	Monandria Diandria, &c. *as in the Classes.*
23. Polygamia	⎰⎱	Monœcia Diœcia Triœcia
24. Cryptogamia	⎰⎱	Filices Musci Hepaticæ Algæ Fungi

Artificial Analysis of the Natural Orders.

Div. I. VASCULARES.
Class 1.—DICOTYLEDONES or EXOGENÆ.

Tribe 1. Angiospermæ.
* Polypetalæ.
  † Thalamifloræ.
  ‡ Apocarpæ.
  ‡‡ Syncarpæ.
¶ *Ovarium many celled: ovula attached to the face of the dissepiments.*
¶ ¶ *Ovarium 1-celled: ovula parietal.*
¶ ¶ ¶ *Ovarium 2- or more-celled: ovula attached to the axis;*
or, *Ovarium 1-celled: ovula adhering to a central placenta.*
  † Calycifloræ.
  ‡ Apocarpæ.
  ‡‡ Syncarpæ.
¶ *Ovarium superior.*
¶ ¶ *Ovarium inferior.*
  ** Apetalæ.
¶ *Ovula indefinite.*
¶ ¶ *Ovula definite.*
  *** Achlamydeæ.
  **** Monopetalæ.
¶ *Ovarium more or less inferior.*
¶ ¶ *Ovarium superior.*
  (a.) *flowers regular.*
  (b.) *flowers irregular.*
Tribe 2. Gymnospermæ.
Class 2.—MONOCOTYLEDONES or ENDOGENÆ.
Tribe 1. Petaloideæ.
  * Tripetaloideæ.
  ** Hexapetaloideæ.
  *** Spadiceæ.
Tribe 2. Glumaceæ.

Div. II. CELLULARES.
or, Class 3.—ACOTYLEDONES.
  ·* Filicoideæ.
  ** Muscoideæ.
  *** Aphyllæ.

6

II. PHYSIOLOGICAL. — *On Mondays, Wednesdays, and Fridays.*

Animals and Vegetables (*organised*); Minerals (*unorganised*).
Attraction, affinity, vitality.
Properties of the vegetable tissue—*extensibility, elasticity, hygroscopicity.*

Physiology.
Epirreology.

The vital property or "excitability" of plants: *irritability—sleep.*

Vegetable functions.

   I. Nutrition (*7 periods*).
     (1.) absorption *(by spongioles.)*
     (2.) ascent of sap.—*(Endosmose.)*
       circulation of chara, &c.
     (3.) exhalation *(by stomata.)*
     (4.) respiration *(in leaf.)*
       Proper juice; gum; fecula; sugar; lignine.
       Etiolation of leaves.
     (5.) Descent and diffusion of nutritive materials.
       Development of buds; Lenticellæ.
     (6.) Assimilation.
       Increase of stem and root *(in length and thickness.)*
         (*a.*) Central system : *(pith, wood, alburnum.)*
         (*b.*) Cortical system : *(liber, bark, epidermis.)*
       Effects of pruning and grafting.
       Duration of plants.
     (7.) Secretions: *excretions.*
       Rotation of Crops.
       Vegetable products; composition, &c.
       Nutrition of Parasites and Epiphytes.

II. Reproduction (5 *periods*).
(1.) flowering (*its periodicity*).
(2.) fertilization (*of ovule*).
Hybrids.
(3.) maturation (*of fruit and seeds*).
(4.) dissemination (*of seeds*).
(5.) germination (*of embryo*).
Propagation by subdivision.
Various modes of reproduction.

Botanical Geography.

ELEMENTARY BOTANICAL WORKS.

* Lindley's Introduction to Botany. 8vo. 1832.
——————————— the Natural System. 8vo. 1830.
——————— Synopsis of the British Flora. 12mo.
Macgillivray's Withering's Botany.
* Hooker's British Flora. 2d edit. 8vo.
* Lloyd's Botanical Terminology.

De Candolle's Theorie Elementaire. 8vo. 1819.
——————— Organographie. 2 vols. 8vo. 1827.
——————— Physiologie vegetale. 3 vols. 8vo. 1832.

BOTANICAL APPARATUS.

Vasculum, large and small.
Chalk-paper: Old Book-covers: Press: Digging-knife:
Forceps, lens, and penknife (*for the lecture-room*).
Microscope.

QUESTIONS

ON THE

SUBJECT-MATTER

OF

SIXTEEN LECTURES IN BOTANY,

REQUIRED FOR A

PASS-EXAMINATION.

BY

Rev. J. S. HENSLOW, M. A.

PROFESSOR OF BOTANY.

CAMBRIDGE:

DEIGHTON; MACMILLAN & CO.

M.DCCC.LI.

PREFACE.

In so extensive a Science as Botany it is necessary to limit, with sufficient precision, the amount to be required as a minimum from those who may wish to attend a course of Lectures, for the purpose of complying with the new regulations concerning degrees. Three years ago I published a Syllabus by way of suggesting what I considered might be fairly required for a Pass-examination. In the hope of yet more definitely fixing this amount, I am now induced to publish the present List of Questions which have reference to a portion of the subject-matter introduced at sixteen of the Lectures delivered during the present term. Whoever may be expecting to acquire a competent knowledge of this subject by merely listening to what shall be told him at lectures, will be disappointed. "How to observe" is an art to be acquired by "observing," and not by listening, or even by reading alone. The Student will find himself confused rather than enlightened if he will not take the trouble to examine plants, and to compare what he sees in them with the descriptions and definitions by which they are to be recognized. If he will consent to do this, he will soon find a growing interest in the subject ; and it is certainly one which need not interfere with the regular course of reading exacted of him for his degree. My advice to all who desire to master so much of Botany as may be required for a Pass-examination, is to attend these Lectures in their first year. In occasional visits to the Botanic Garden, during walks in the country, and especially in the long vacation, ample opportunities will be found for acquiring far more than will be necessary for this object. It must be distinctly understood that responses to the questions here proposed, are not to be confined to matter that may be acquired by rote : I trust no one will attempt such an uninstructive and laborious method of proceeding. Students will be expected to shew that they have learnt how to apply the several terms to which reference is here made, to the particular objects which they are designed to illustrate. For instance, one question asked is to " Distinguish between an indefinite and a definite inflorescence," and the obvious reply to this question may be further checked by reference to some half dozen specimens on the table, and the Student asked to point out which of them will satisfy one or other of the conditions specified in his own definition. Again, a certain number of Genera have been named, from which some species are to be selected for " Demonstrations " according to a given formula. But it will not be required that a Student should frame a reply to such parts of this formula as the specimens before him do not afford an opportunity for " seeing " how he ought to do so. For instance, if a specimen of some Ranunculus have no fruit upon it, he will not be required to describe the fruit and seed according to the formula. And yet any one who may have studied the fruit of a few species of this genus, with a view to the examinations, will probably be able to recollect that all have " Achenes," containing "one erect seed," which is "albuminous," and that it has a minute "homotropous embryo;" or, the last character might be inferred from finding the ovary to contain an "anatropous ovule." Whoever may happen to remember any such general characters of the few genera that have been selected, will do well to state them, and he will be allowed credit for such knowledge, even though the facts detailed do not appear upon the specimen under examination. It is always advisable to commit to memory a few of these important characteristics of the typical genera in the larger Orders.

Many will rejoice in after life that the University has required them to pass an examination in one or other of the subjects treated in our Professorial Lectures. By mastering, whilst at College, the little difficulties which stand in the way of all who wish to learn "how to observe" the works of nature, they will have obtained a sufficient grasp of such subjects to make them afterwards agreeable as well as instructive occupations. If they will trust the experience of one who has passed the last twelve years in an out of the way purely agricultural country village, where even many a learned classic or profound mathematician might very possibly have felt himself occasionally at a loss for want of further intellectual companionship than books alone can afford, I can assure them there is nothing like possessing the power of conversing with the works of nature for a constantly unalloyed source of enjoyment at all seasons of the year. However the wit or the wisdom of man may delight us, the excellency of God's works leaves them far behind. I may be presumptuous in making the remark, and many will esteem it ill-advised ; but it has often appeared to me, that the study of God's Word alone may not always be sufficient to protect some of us against the morbid imaginings of these uncertain days ; and I therefore the more rejoice to find the Natural Sciences at length taking firmer root in this seat of sound learning and religious education than they have hitherto obtained.

J. S. HENSLOW.

CAMBRIDGE,
May 30, 1851.

QUESTIONS on the Subject Matter introduced at the BOTANICAL LECTURES during the EASTER TERM, 1851.

N.B. The Pages referred to are those of the Syllabus.

LECTURE I.

p. 2. (1) Name the " Conservative Organs" of plants, and give the reason why they are so termed.

— (2) The same for the " Reproductive Organs."

p. 7. (3) Under how many separate groups (at least) is every plant systematically arranged; stating the subordination of the lower to the higher?

— (4) Under what subordinate " groups" may the individuals of a given "species" be collected?

LECTURE II.

p. 2. (5) Define the " Floral Whorls," and the " Organs" which constitute the subordinate parts of each whorl.

— (6) Distinguish those organs (or any part of them) among the Floral Whorls, which are generally essential, from those which are not essential to the reproductive functions of the flower.

p. 2. (7) To what considerations do we refer the "modifications" of the floral organs, from whence result the immense diversity of character noticeable in the flowers of plants?

— (8) Which are the parts in a flower that are "essentially" changed into parts of the fruit; and what fresh names do those parts acquire when such change has been effected.

— (9) From what parts of the flower have the peculiar "Exuvial Appendages" originated, which we find about the fruit of

Castanea Vesca (Chestnut),
Mirabilis Jalapa (Marvel of Peru),
Physalis Alkakengi (Winter Cherry)?

— (10) Which of the following seeds are "albuminous" and which "ex-albuminous?" and what are the parts distinguishable in their Embryos?

Vicia Faba (Garden Bean),
Ricinus Communis (Castor-oil Plant),
Arum maculatum (Spotted Arum),
Zea Mays (Indian Corn).

— (11) Distinguish between the parts which give a false resemblance to

Æsculus hippocastanum (Horse Chestnut) seed, and
Castanea vulgaris (Chestnut) fruit.

— (12) What is the peculiar character which distinguishes an "acotyledonous spore or sporule" from a "cotyledonous seed?"

LECTURE III.

p. 3. (13) Distinguish between a "definite" and an "indefinite" stem.

— (14) What is meant by a "node," and what by an "internode?"

— (15) Be prepared to define the terms "Suffrutescent," "Rhizome," "Corm," "Bulb," "Tuber," "Thorn," "Prickle."

p. 19. (16) Demonstrate some specimen of Primula, Hottonia, Anagallis, or Samolus.

N.B. In this and all subsequent "Demonstrations," so much of the following Formula is to be used as the specimens may afford an opportunity for noticing the characters here referred to.

FORMULA FOR DEMONSTRATION.

I.   Root.   General Character.

II.   Stem.   General Character.

III.   Leaves.

       (1) Position.

       (2) Arrangement.

       (3) Insertion.

       (4) Stipulation.

       (5) Composition.

       (6) Form.

       (7) Margin and Incision.

       (8) Surface.

       (9) Venation.

10

IV. INFLORESCENCE.
    (1) General Character.
    (2) Bracteal Appendages.

V. FLOWER: Peculiarities in the number, form, and arrangement of its organs.
    (1) Insertion of Corolla and Stamens
    (2) Perianth.
      (a) Calyx.
      (b) Corolla.
    (3) Andrœcium.
    (4) Gynœcium.
      (a) Ovary.
      (b) Style.
      (c) Stigma.
      (d) Ovules.
    (5) Nectariferous Appendages.

VI. FRUIT.
    (1) General Character.
    (2) Exuvial Appendages.
    (3) Dehiscence.
    (4) Placentation.

VII. SEED. General Character.

VIII. EMBRYO. Structure; or reasons for inferring this if indistinct.

IX. PLANT.
    (1) General Character.
    (2) Linnean System. Actual position determined; and any reason for considering it misplaced there.

(3) Natural System. The Class, Sub-class (if any), and Order to which it may be referred; or to which it seems to approximate.

— (17) In what respect do Hottonia, Anagallis, and Samolus, chiefly deviate from the normal type of Primulaceæ, as exemplified by the genus Primula?

LECTURE IV.

p. 3. (18) Be prepared to define the terms " Petiole," " Pinnate," " Palmate," " Pedate," " Peltate," " Decurrent," " Perfoliate."

— (19) Distinguish between an "alternate," an "opposite," and a "verticillate" arrangement of leaves.

— (20) What is a " Phyllodium?"

p. 7. (21) What are the most usual characteristics by which we can discriminate between the leaves of Dicotyledons and Monocotyledons?

p. 13. (22) Demonstrate some specimen of Viola.

p. 8. (23) What are the characters by which the first ten classes of the Linnean System are distinguished, and by what are their Orders determined?

LECTURE V.

p. 3. (24) Distinguish between an "Indefinite" and a "Definite" Inflorescence.

— (25) Be prepared to define the terms "Spike," "Raceme or Cluster," "Corymb," "Umbel," "Head," "Cyme."

p. 11. (26) Demonstrate some specimen of Clematis, Anemone, Ranunculus, Caltha, Helleborus, Aquilegia, or Pœonia.

— (27) What are the three chief modifications observable in the character of the fruit of Ranunculaceæ?

— (28) What are the characters upon which the Linnean Classes Dodecandria, Icosandria, and Polyandria, and their orders, depend? How do we readily distinguish between the two latter?

LECTURE VI.

p. 4. (29) Distinguish between a Valvate and an Imbricate Æstivation.

— (30) What part of the Receptacle is the Disk?

— (31) Distinguish between Dichlamydeous, Monochlamydeous, and Achlamydeous Flowers

13

p. 4. (32) Explain the terms Hypogynous, Perigynous and Epigynous.

p. 21. (33) Describe the Inflorescence of Arum and Lemna.

p. 12. (34) Demonstrate some specimen of Cruciferæ.

— (35) Describe the five principal modifications in the Embryos of Cruciferæ: and the six conditions of their pericarps, upon which the Suborders and Tribes of this Order are founded.

p. 8. (36) What are the characters upon which the Linnean Classes Didynamia and Tetradynamia, and their respective orders, depend?

LECTURE VII.

p. 5. (37) Distinguish between a " septicidal" and a " loculicidal" dehiscence.

— (38) There are (apparently) eight stamens in the terminal flower, and ten in the lateral flowers of the inflorescence in Adoxa Moschatellina. To which of the Linnean Classes would you refer for this plant? As the stamens appear to alternate in pairs with the divisions of the corolla, and each bears a single-celled anther, what may we infer with respect to their real structure?

14

p. 14. (39) Demonstrate some specimen of Genista, Sarothamnus, Medicago, Vicia or Lathyrus.

— (40) Define a "Papilionaceous" flower; and state the three several conditions (in respect of cohesion of their filaments) under which the stamens of such flowers occur?

— (41) Explain how the Legume of an Astragalus becomes spuriously two-celled.

— (42) Explain the peculiarity of the Legumes termed "lomentaceous."

p. 8. (43) What are the characters upon which the Linnean Classes Monadelphia, Diadelphia, Polyadelphia, and their respective Orders, depend?

LECTURE VIII.

p. 5. (44) Explain the nature of the hilum, chalaze, raphe, and micropyle.

— (45) What is the origin of the true "Arillus," as distinguishable from the "Arillode" or false Arillus.

— (46) Distinguish between the Orthotropous, Campylotropous, and Anatropous condition of ovules.

— (47) Distinguish between the Antitropous, Amphitropous, Homotropous, and Heterotropous direction of embryos.

p. 20 (48) The floral organs in Paris quadrifolia affect a quaternary arrangement ; the perianth might easily be esteemed "dichlamydeous," the veins of the well-developed leaves are branched. Without examining the seed, how would you be able to conclude (notwithstanding these characters of the flower and leaves) that this plant is monocotyledonous?

p. 14 (49) Demonstrate some specimen of Prunus, Geum, Fragaria, Poterium, Cratægus or Pyrus.

— (50) Be prepared to explain the characters of any of the five following forms of fruit found among Rosaceæ—
Drupe, Follicle, Æterio, Hip, Pome.

LECTURE IX.

p. 14. (51) Demonstrate some specimen of Geranium or Erodium.

p. 15. (52) Demonstrate some specimen of Apargia, Sonchus, Leontodon, Carduus (including Cnicus), Centaurea, Bellis, Cineraria, Senecio, or Petasites.

p. 16. (53) Explain the origin of "Pappus," in Compositæ, and distinguish between the "stipitate" and "sessile," the "pilose" and "plumose."

p. 16. (54) How are the three great groups "Corymbiferæ," "Cynerocephalæ," and "Cichoraceæ" (considered as Orders by Jussieu) characterised? Which of these is the most natural, and to be retained as a sub-Order, and under what other sub-Orders can we dispose of the species included in the other two. Explain the conditions under which the capitula occur in each case.

— (55) On what character is the Linnean Class Syngenesia framed; and what are the conditions of the florets in its five Orders, "Æqualis," "Superflua," "Frustranea," "Necessaria," "Segregata"?

— (56) Distinguish between "Homogamous" and "Heterogamous" Capitula; and between "Monoicous" and "Heterocephalous" species.

p. 17. (57) How do Valerianaceæ and Dipsaceæ chiefly resemble Compositæ; and by what prominent differences are they readily separable from this extensive Order?

LECTURE X.

p. 6. (58) To what single condition may we refer the variously modified "Elementary Organs" of which plants are composed, and how do these modifications apparently originate?

p. 2. (59) Distinguish between the "Spiral," "Dotted," "Annular," "Reticulated," and "Scalariform" structure of the Elementary Organs.

— (60) Explain the structure of Epidermis; and the manner in which pubescence occurs on its surface.

p. 20. (61) Demonstrate some specimen of Fritillaria, Agraphis, or Allium.

p. 18. (62) Demonstrate some specimen of Borago, Symphytum, or Pulmonaria.

— (63) Demonstrate some specimen of Salvia, Lamium, or Ajuga.

— (64) Explain the nature of the "Inflorescence," and "Fruit" of Labiatæ.

— (65) Are there any Labiatæ which are not comprehended in the Linnean system under Didynamia Gymnospermia; or any of the latter which do not belong to Labiatæ?

LECTURE XI.

p. 6. (66) Compare the internal structure of Exogens, Endogens, and Acrogens.

p. 13. (67) Demonstrate some specimen of Reseda; and mention the peculiarity noticeable in the capsule of this genus.

p. 15. (68) Demonstrate some specimen of Sanicula,
Æthusa, Smyrnium, Scandix, Anthriscus,
Chærophyllum, or Myrrhis.

— (69) Be prepared to define the terms " Carpo-
phorus," " Cremocarp," " Commissure,"
" Ridge," " Interstice," " Vitta," as applied to the fruit of Umbelliferæ.

— (70) What is the most usual condition of the
" Infloresence " of Umbelliferæ, and to
what modifications is it subject ?

LECTURE XII.

p. 7. (71) Explain the anatomy of the leaf.

p. 23. (72) In what respect are the conditions under
which " Organic compounds " originate
essentially distinguishable from those under which crystalline aggregations may
take place ?

p. 23. (73) What elements enter into the composition
of " Non-nitrogenous" and " Nitrogenous"
organic compounds; and what are the
proportions to which their equivalents approximate in Starch and Albumen respec-
tively ?

p. 13. (74) Explain the structure of the flower in
Polygala.

p. 19. (75) Demonstrate some specimen of Euphorbia
or Mercurialis.

pp. 11 to 19. (76) Upon what considerations have
the names " Thalamifloræ," " Calycifloræ,"
" Corollifloræ," and " Monochlamydeæ,"
been given to subordinate groups, considered as four sub-classes of Dicotyledones?

LECTURE XIII.

p. 18. (77) Demonstrate some specimen of Hyoscyamus or Solanum.

— (78) Demonstrate some specimen of Veronica,
Scrophularia, Antirrhinum, Linaria, or Verbascum.

— (79) Under which of the Linnean Classes are
different Genera of Scrophulariaceæ to be
found arranged?

— (80) Mention the principal distinguishing cha-
racteristic between Solanaceæ and Scrophulariaceæ.

p. 21. (81) Demonstrate some specimen of Orchis,
Aceras, or Ophrys.

— (82) Be prepared to define the terms " Rostel-
lum," " Bursicula," " Retinaculum," and
" Caudicula," as applied to the description
of Orchidaceæ.

p. 8. (83) What are the characters on which the Linnean class Gynandria and its Orders depend?

p. 23. (84) Distinguish between the causes to which we may ascribe the rapid movements noticeable in the ripened stamens of the Nettle, and of the Barberry.

LECTURE XIV.

p. 20. (85) Demonstrate some specimen of Iris.

p. 19. (86) Describe the Inflorescence and Flowers of some species of Salix or Populus; and the fruit and seed of some Quercus, Corylus, or Castanea.

p. 20. (87) What is the peculiarity characterising the carpellary arrangement of Coniferæ? and what are the parts noticeable in the fruit of Pinus and Taxus?

p. 8. (88) On what characters are the Linnean Classes Monœcia, Diœcia, Polygamia, and their respective Orders, founded?

p. 24. (89) By what specific processes is the function of nutrition carried on?

LECTURE XV.

p. 21. (90) What are the subordinate parts in the Inflorescence and in the Flowers of Gramineæ?

— (91) Demonstrate some specimen, either in flower or in fruit, of our common Cerealia of the genera Triticum, Hordeum, Avena, or Secale.

p. 22. (92) Demonstrate some specimen of Carex.

p. 25. (93) Under what conditions, and by what special organs, are the so-called processes of " Exhalation," and " Respiration," carried on? and what is the immediate result produced?

LECTURE XVI.

p. 10. (94) In determining the natural affinities of plants, what is the subordination of the one to the other of those reproductive organs from which characters of different values are to be derived?

p. 28. (95) In what state are the ovules at the period of flowering; describing their several parts by the technical terms applied to them.

— (96) Explain the process by which the fertilization of the ovules is secured.

p. 28. (97) What are the changes induced during the progress of "maturation," from the first appearance of the embryo to the ripened seed?

———————————

Hooker, William Jackson: *Icones Plantarum* (10 vols., 1837–54)
[ISBN 9781108039314]

Hooker, William Jackson: *Kew Gardens* (1858) [ISBN 9781108065450]

Jussieu, Adrien de, edited by J.H. Wilson: *The Elements of Botany* (1849)
[ISBN 9781108037310]

Lindley, John: *Flora Medica* (1838) [ISBN 9781108038454]

Müller, Ferdinand von, edited by William Woolls: *Plants of New South Wales* (1885)
[ISBN 9781108021050]

Oliver, Daniel: *First Book of Indian Botany* (1869) [ISBN 9781108055628]

Pearson, H.H.W., edited by A.C. Seward: *Gnetales* (1929) [ISBN 9781108013987]

Perring, Franklyn Hugh et al.: *A Flora of Cambridgeshire* (1964)
[ISBN 9781108002400]

Sachs, Julius, edited and translated by Alfred Bennett, assisted by W.T. Thiselton Dyer:
A Text-Book of Botany (1875) [ISBN 9781108038324]

Seward, A.C.: *Fossil Plants* (4 vols., 1898–1919) [ISBN 9781108015998]

Tansley, A.G.: *Types of British Vegetation* (1911) [ISBN 9781108045063]

Traill, Catherine Parr Strickland, illustrated by Agnes FitzGibbon Chamberlin:
Studies of Plant Life in Canada (1885) [ISBN 9781108033756]

Tristram, Henry Baker: *The Fauna and Flora of Palestine* (1884)
[ISBN 9781108042048]

Vogel, Theodore, edited by William Jackson Hooker: *Niger Flora* (1849)
[ISBN 9781108030380]

West, G.S.: *Algae* (1916) [ISBN 9781108013222]

Woods, Joseph: *The Tourist's Flora* (1850) [ISBN 9781108062466]

For a complete list of titles in the Cambridge Library Collection please visit:
http://www.cambridge.org/features/CambridgeLibraryCollection/books.htm

Selected botanical reference works available in the
CAMBRIDGE LIBRARY COLLECTION

al-Shirazi, Noureddeen Mohammed Abdullah (compiler), translated by Francis Gladwin: *Ulfáz Udwiyeh, or the Materia Medica* (1793) [ISBN 9781108056090]

Arber, Agnes: *Herbals: Their Origin and Evolution* (1938) [ISBN 9781108016711]

Arber, Agnes: *Monocotyledons* (1925) [ISBN 9781108013208]

Arber, Agnes: *The Gramineae* (1934) [ISBN 9781108017312]

Arber, Agnes: *Water Plants* (1920) [ISBN 9781108017329]

Bower, F.O.: *The Ferns (Filicales)* (3 vols., 1923–8) [ISBN 9781108013192]

Candolle, Augustin Pyramus de, and Sprengel, Kurt: *Elements of the Philosophy of Plants* (1821) [ISBN 9781108037464]

Cheeseman, Thomas Frederick: *Manual of the New Zealand Flora* (2 vols., 1906) [ISBN 9781108037525]

Cockayne, Leonard: *The Vegetation of New Zealand* (1928) [ISBN 9781108032384]

Cunningham, Robert O.: *Notes on the Natural History of the Strait of Magellan and West Coast of Patagonia* (1871) [ISBN 9781108041850]

Gwynne-Vaughan, Helen: *Fungi* (1922) [ISBN 9781108013215]

Henslow, John Stevens: *A Catalogue of British Plants Arranged According to the Natural System* (1829) [ISBN 9781108061728]

Henslow, John Stevens: *A Dictionary of Botanical Terms* (1856) [ISBN 9781108001311]

Henslow, John Stevens: *Flora of Suffolk* (1860) [ISBN 9781108055673]

Henslow, John Stevens: *The Principles of Descriptive and Physiological Botany* (1835) [ISBN 9781108001861]

Hogg, Robert: *The British Pomology* (1851) [ISBN 9781108039444]

Hooker, Joseph Dalton, and Thomson, Thomas: *Flora Indica* (1855) [ISBN 9781108037495]

Hooker, Joseph Dalton: *Handbook of the New Zealand Flora* (2 vols., 1864–7) [ISBN 9781108030410]

CAMBRIDGE
UNIVERSITY PRESS

University Printing House, Cambridge, CB2 8BS, United Kingdom

Published in the United States of America by Cambridge University Press, New York

Cambridge University Press is part of the University of Cambridge.
It furthers the University's mission by disseminating knowledge in the pursuit of
education, learning and research at the highest international levels of excellence.

www.cambridge.org
Information on this title: www.cambridge.org/9781108061728

© in this compilation Cambridge University Press 2013

This edition first published 1829
This digitally printed version 2013

ISBN 978-1-108-06172-8 Paperback

This book reproduces the text of the original edition. The content and language reflect
the beliefs, practices and terminology of their time, and have not been updated.

Cambridge University Press wishes to make clear that the book, unless originally published
by Cambridge, is not being republished by, in association or collaboration with, or
with the endorsement or approval of, the original publisher or its successors in title.

A Catalogue of

British Plants

Arranged According to the Natural System

With the Synonyms of
De Candolle, Smith, and Lindley

John Stevens Henslow

Cambridge University Press has long been a pioneer in the reissuing of out-of-print titles from its own backlist, producing digital reprints of books that are still sought after by scholars and students but could not be reprinted economically using traditional technology. The Cambridge Library Collection extends this activity to a wider range of books which are still of importance to researchers and professionals, either for the source material they contain, or as landmarks in the history of their academic discipline.

Drawing from the world-renowned collections in the Cambridge University Library and other partner libraries, and guided by the advice of experts in each subject area, Cambridge University Press is using state-of-the-art scanning machines in its own Printing House to capture the content of each book selected for inclusion. The files are processed to give a consistently clear, crisp image, and the books finished to the high quality standard for which the Press is recognised around the world. The latest print-on-demand technology ensures that the books will remain available indefinitely, and that orders for single or multiple copies can quickly be supplied.

The Cambridge Library Collection brings back to life books of enduring scholarly value (including out-of-copyright works originally issued by other publishers) across a wide range of disciplines in the humanities and social sciences and in science and technology.

CAMBRIDGE LIBRARY COLLECTION

Books of enduring scholarly value

Life Sciences

Until the nineteenth century, the various subjects now known as the life sciences were regarded either as arcane studies which had little impact on ordinary daily life, or as a genteel hobby for the leisured classes. The increasing academic rigour and systematisation brought to the study of botany, zoology and other disciplines, and their adoption in university curricula, are reflected in the books reissued in this series.

A Catalogue of British Plants
Arranged According to the Natural System

In 1829, botany had much to prove. A prominent lecturer, John Lindley, noted that 'it has been very much the fashion of late years, in this country, to undervalue the importance of this science, and to consider it an amusement for ladies rather than an occupation for the serious thoughts of man'. In the three documents reissued here, Cambridge botany professor John Stevens Henslow (1796–1861) demonstrates the exacting standards of his course. The work contains an 1829 catalogue of British plants, the skeleton structure of sixteen lectures for 1833 and an 1851 list of potential examination questions. Students were expected to differentiate between 'an indefinite and a definite inflorescence', to recognise 'albuminous seeds', and describe 'nectariferous appendages'. With a strongly Linnaean approach to taxonomy, this collection offers researchers a window into the growth of academic botany prior to the revolution occasioned by Henslow's pupil, Charles Darwin.